Cantilever beam $\Delta = \dfrac{L^3}{3EI}$

fixed at both end $= \Delta = \dfrac{L^3}{12EI}$

SEISMIC DESIGN

for the

Civil

Professional Engineering Examination

Third Edition

Michael R. Lindeburg, P.E.

PROFESSIONAL ENGINEERING REGISTRATION PROGRAM

San Carlos, CA 94070

ACKNOWLEDGMENT

The questions appearing on pages 79-87 are reprinted or reconstructed from previous year's California P.E. examinations with permission by the Board of Registration for Professional Engineers and Land Surveyors. These examinations are copyrighted by the Board of Registration for Professional Engineers and Land Surveyors. Complete examinations may be available from the Board at a nominal cost.

In the *ENGINEERING REVIEW MANUAL SERIES*

Engineer-In-Training Review Manual
Quick Reference Summary Cards for the E-I-T Exam
Mini-Exams for the E-I-T Exam
Civil Engineering Review Manual
Seismic Design for the Civil P.E. Exam
Timber Design for the Civil P.E. Exam
Structural Engineering Practice Problem Manual
Mechanical Engineering Review Manual
Electrical Engineering Review Manual
Chemical Engineering Review Manual
Chemical Engineering Practice Exam Set
Land Surveyor Reference Manual
Expanded Interest Tables
Engineering Law, Design Liability, and Professional Ethics

In the *ENGINEERING CAREER ADVANCEMENT SERIES*

How To Become a Professional Engineer
The EXPERT WITNESS Handbook —A Guide for Engineers—
Getting Started as a Consulting Engineer
A Company Policy and Personel Handbook

Distributed by: Professional Publications, Inc.
Post Office Box 199
Department 77
San Carlos, CA 94070
(415) 593-9119

**SEISMIC DESIGN for the Civil Professional Engineering Exam
3rd edition**

Printed in the United States of America

Library of Congress number: 80-81796

ISBN: 0-932276-32-6

Professional Engineering Registration Program
Post Office Box 911, San Carlos, CA 94070

Current printing of this edition (last number) 10   9   8   7   6

PROFESSIONAL ENGINEERING REGISTRATION PROGRAM
P.O. Box 911, San Carlos, CA 94070

TABLE OF CONTENTS

PROFESSIONAL ENGINEERING REGISTRATION PROGRAM
P.O. Box 911, San Carlos, CA 94070

INTRODUCTION

This document is designed to be your primary reference for answering earthquake problems on the civil engineering exam.

The trend in the format of earthquake problems has been towards analysis and away from commentary. Although problems from the early 1970's were largely definition-oriented, recent problems have required examinees to display computational abilities.

None of the analytical techniques required in the earthquake problems are proprietary or obscure. This document, with its many examples and explanations, presents these techniques in step-by-step formats. The explanations are a blend of practical information and simple vibration theory. Complex proofs and derivations have been omitted, since they are not relevant to the examination.

In addition to a detailed index, set of definitions, and complete listing of nomenclature, this document contains typical examination problems. Although there is no guarantee that any problem type presented will appear on future examinations, it is suggested that you become familiar with this material by using the index to answer the typical problems.

Appreciation is extended to the Structural Engineering Association of California for permission to reprint Section 1 of its "Recommended Lateral Force Requirements."

1. Nomenclature

a	acceleration	ft/sec^2
A	seismogram amplitude, or area	mm, or ft^2
A_O	seismograph reference amplitude	mm
B	damping coefficient, or seismic parameter	lb-sec/ft, or -
C	base shear coefficient	-
C_F	lateral force coefficient	-
C_T	mode period constant	-
d	distance from epicenter, or wall length	ft
D	fault offset	cm
e	eccentricity, or energy	ft, or ergs
E	modulus of elasticity, or expected number of earthquakes	lb/ft^2, or -
f	linear frequency	hertz
F	force	lb
$F(t)$	forcing function	lb
g	acceleration due to gravity (32.2)	ft/sec^2
G	shear modulus	lb/ft^2
h	height	ft
I	moment of inertia, or occupancy importance coefficient	ft^4, or -
J	polar moment of inertia	ft^4
k	spring stiffness constant	lb/ft
K	construction type coefficient	-
L	girder length, or fault length	ft, or km
m	mass	slugs
M	moment, or Richter magnitude	ft-lb, or -
MF	magnification factor	-
N_O	seismic area parameter	1/(mile)^2
P	amplitude of forcing function	lb
r	distance from zero stress point	ft
R	relative ridigity	lb/in
S	site-structure resonance factor	-
S_a	spectral acceleration	ft/sec^2
S_d	spectral displacement	ft
S_v	spectral velocity	ft/sec

t	time, or thickness	sec, or ft
T	period	sec
v	velocity	ft/sec
V	base shear	lb
W	weight	lb
x	position	ft
Y	number of years	years
Z	region coefficient	-

Symbols

ω	angular frequency	rad/sec
ξ	damping ratio	-
τ	time	sec
Δ	total displacement	ft
δ	mode displacement	ft
ϕ	mode shape factor	-
Γ	modal participation factor	-

Subscripts

c	column
d	damped
e	effective
f	forced
g	ground, or girder
n	nth cycle
o	original
p	p-wave
s	s-wave
t	top

2. Conversions

Multiply	By	To Obtain
ergs	.7376 EE-7	ft-lb
ft-lb	1.356 EE7	ergs
ft/sec^2	.03106	gravities
gravities	32.2	ft/sec^2
gravities	386	in/sec^2
in/sec^2	.002591	gravities
inches	25.4	mm
km	.6214	miles
miles	1.609	km
mm	.03937	inches

3. Definitions

BASE: the level at which the earthquake motions are considered to be imparted to the structure or the level at which the structure as a dynamic vibrator is supported.

BOX SYSTEM: a structural system without a complete vertical load carrying space frame. In this system, the required lateral forces are resisted by shear walls or braced frames.

BRACED FRAME: a truss system or its equivalent which is provided to resist lateral forces and in which the members are subjected primarily to axial stresses.

CONFINED CONCRETE: Concrete stressed in all directions by closely spaced ties which restrain it in directions perpendicular to the applied stress.

CRITICAL DAMPING: that amount of damping which will return the system from initial deformation to zero without reversal.

DAMPING: loss of energy absorbed by internal strains

DIP: the angle that a stratum, joint, fault, or other plane makes with a horizontal plane.

DIP SLIP: the component of the slip parallel with the dip of the fault.

DUCTILE MOMENT RESISTING FRAME: a space frame made ductile by special reinforcement of failure points, such as joints. Subsequent failure at locations of low stress enables the structure to utilize its reserve strength beyond first yield.

ELASTIC REBOUND THEORY: seismic theory based on the tectonic plate concept. It proposes that stresses are created in fault lines by the shifting of the tectonic plates; but the faults will resist and not move until the accumulated stresses overcome internal friction resistance.

EQUIVALENT STATIC LOAD: method of seismic force analysis used by UBC.

ESSENTIAL FACILITIES: Legal term used in codes to define a facility which must remain functional after a disaster such as a major earthquake.

FAULT: a fracture or fracture zone along which the two sides have been displaced relative to one another parallel to the fracture.

FAULT CREEP: apparently continuous displacement along a fault at a low but varying rate, usually not accompanied by felt earthquakes. Fault creep is not necessarily tectonic in origin; it may result from artificial withdrawal of fluids or solids.

FAULT DISPLACEMENT: relative movement of the two sides of fault, measured in any specific direction.

FAULT GOUGE: the filler material which forms between two plates which are sliding aginst each other.

FAULT SAG: a narrow tectonic depression common in strike-slip fault zones. Fault sags are generally closed depressions less than a few hundred feet wide and approximately parallel to the fault zone. Those that contain water are called sag ponds.

FAULT SCARP: a cliff or steep slope formed by displacement of the ground surface.

FRACTURE: a general term for discontinuities in rock. It includes faults, joints, and other breaks.

GOUGE: see "fault gouge."

GRABEN: a fault block, generally long and narrow, that has been dropped down relative to the adjacent blocks by movement along the bounding faults. The same word is used for both the singular and plural.

HERTZ: same as cycles per second

HOOP: one-piece closed tie or continuously wound tie that encloses longitudinal reinforcement.

HYPOCENTER: the actual location of the earthquake beneath the Earth's surface.

IGNEOUS ROCK: rocks formed by the solidification of molten magma.

LATERAL FORCE RESISTING SYSTEM: that part of the structural system designed to resist the lateral forces.

LEFT SLIP: strike-slip displacement in which the block across the fault from an observer has moved to the left.

LIQUIFACTION: the loss of load-carrying ability of loose fill (such as sand) during an earthquake.

MOMENT RESISTING SPACE FRAME: a vertical load carrying space frame in which the members and joints are capable of resisting forces primarily by flexure.

NORMAL FAULT: a fault in which the block above an inclined surface has moved downward relative to the block below the fault surface; also includes vertical faults with vertical slip.

OBLIQUE SLIP: a combination of strike slip and normal or reverse slip.

PARAPET: a low wall or railing protecting a ledge. The term is generally used to describe decorative railings on roof tops.

PLASTIC HINGE: region where ultimate moment strength of a member may be developed and maintained with corresponding significant inelastic rotation as primary tension reinforcement elongates beyond yield strain.

RESERVE ENERGY: energy which a ductile system is capable of absorbing by plastic strains.

RESONANCE: the condition when the frequency of excitation equals the frequency of the building response.

REVERSE FAULT: a fault in which the block above an inclined fault surface has moved upward relative to the block below the fault surface.

RIGHT-NORMAL SLIP: fault displacement consisting of nearly equal components of right slip and normal slip.

RIGHT SLIP: strike-slip displacement in which the block across the fault from an observer has moved to the right.

RIGID FRAME: a vertical load carrying frame where the members and joints resist forces by rotation and flexture.

SHEAR WALL: a wall designed to resist lateral forces parallel to the plane of the wall.

SLIP: the relative displacement of points on opposite sides of a fault, measured on the fault surface.

SPACE FRAME: a three-dimensional structural system without bearing walls, composed of interconnected members laterally supported so as to function as a complete self-contained unit with or without the aid of horizontal diaphragms or floor bracing systems.

SPECIAL DUCTILE FRAME: a structural frame composed of reinforced concrete flexural members and columns with cast-in-place connections designed and detailed to accommodate reversible lateral displacements after formation of plastic hinges.

SPECIAL SHEARWALL: reinforced concrete shearwall designed and detailed in accordance with ACI 318 appendix A.

STIRRUP-TIE: closed stirrup conforming to definition of a hoop.

STRIKE SLIP: the component of the slip parallel with the strike of the fault; the horizontal component of slip.

STRIKE: the direction or bearing of a horizontal line in the plane of an inclined or vertical stratum, joint, fault, or other plane.

STRIKE-SLIP FAULT: a fault in which the slip is approximately in the direction of the strike of the fault; also called wrench or transcurrent fault.

SUPPLEMENTARY CROSSTIE: tie with a standard 180-deg hook at each end.

TECTONIC: of, pertaining to, or designating the rock structure and external forms resulting from deep-seated crustal and subcrustal forces in the earth.

TECTONIC CREEP: fault creep of tectonic origin; also called slippage.

4. The Nature of Earthquakes

A. What Earthquakes Are

An earthquake is the oscillatory, sometimes violent movement of the Earth's surface that follows a release of energy in the crust. This energy can be generated by a sudden dislocation of segments of the crust, by a volcanic eruption, or even by man-made explosions. Most of the destructive quakes, however, are caused by dislocations of the crust.

When subjected to deep-seated forces (whose origins and natures are largely unknown) the crust may first bend and then, when the stress exceeds the strength of the rocks, break and "snap" to a new position. In the process of breaking, vibrations called "seismic waves" are generated. These waves travel from the source of the earthquake to more distant places along the surface and through the Earth at varying speeds depending on the medium through which they move. Some of the vibrations are of high enough frequency to be audible, while other are of very low frequency with a period of many seconds or minutes.

Most earthquakes occur in areas bordering the Pacific Ocean. This circum-Pacific belt, called the "ring of fire," includes the Pacific coasts of North and South America, the Aleutians, Japan, Southeast Asia, and Australia.

The United States has experienced less destruction than other countries located in this earthquake zone, but millions of Americans live in potential earthquake areas. Large parts of the western United States are known to be particularly vulnerable. Nuclear reactors, great dams, schools, high-rise apartments, and other housing developments are being planned and built in places where there is a high-probability of major earthquakes. This has created an urgent need for more information on the nature, causes, and effects of earthquakes.

Figure 1

Major Earthquake Zones

World map showing the location of earthquakes of magnitude 8.0 and higher, 1897-1976. These great earthquakes occur primarily along convergent or strike-slip plate boundaries and in zones of compressive or strike-slip deformation within continents.

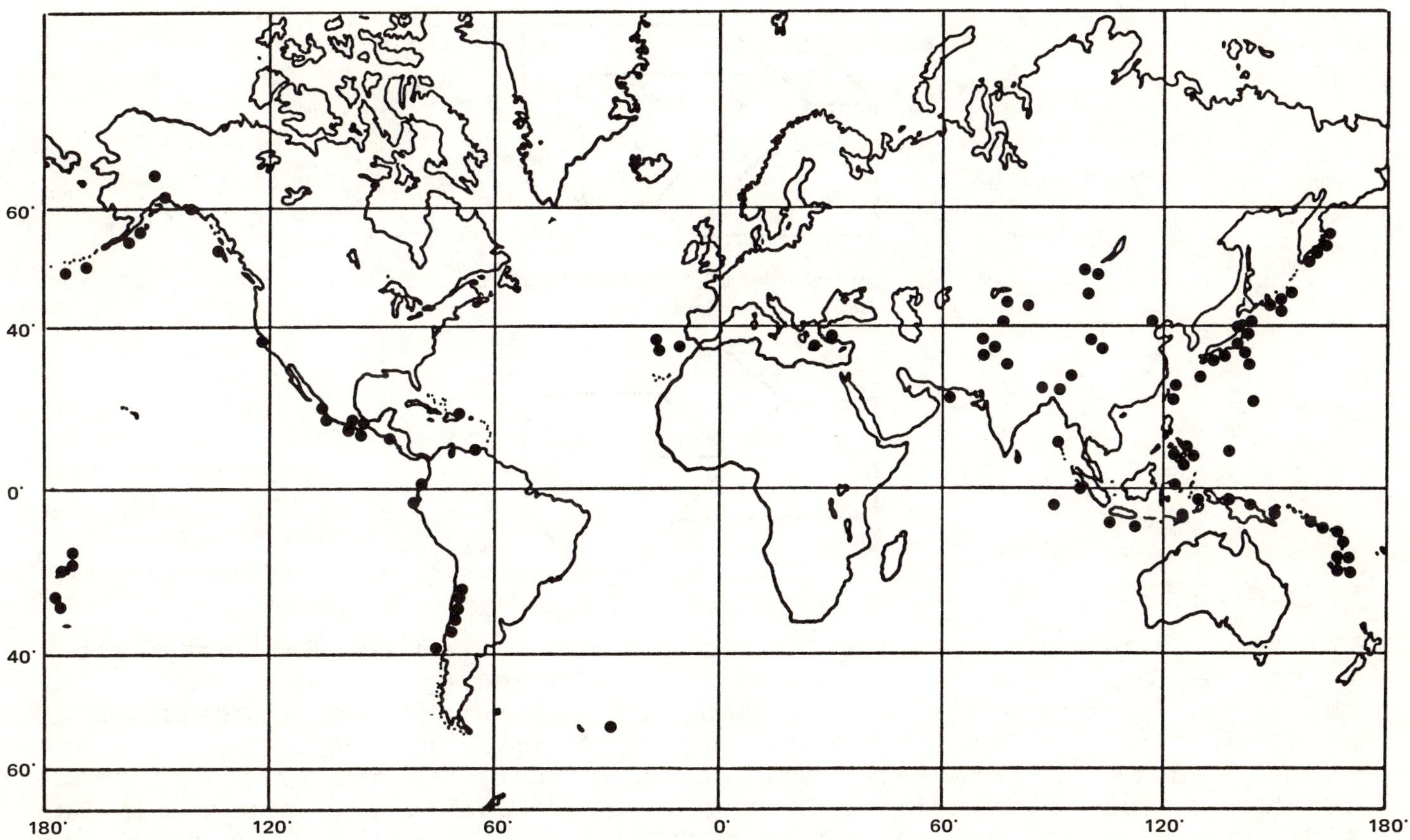

B. Faults

A fault is a fracture in the Earth's crust along which two blocks of the crust have slipped with respect to each other. One crustal block may move horizontally in one direction while the block facing it moves in the opposite direction. Or, one block may move upward while the other moves downward.

Faults are distinguished by the kinds of movements that characterize them. Movement along California's famous San Andreas fault is predominatly horizontal. The fault is called a strike-slip or lateral fault. A fault in which the movement is vertical is called a dip-slip or normal fault. Along many faults, movement is both horizontal and vertical.

Geologists have found that earthquakes tend to reoccur along faults which reflect zones of weakness in the Earth's crust. The fact that a fault zone has recently experienced an earthquake offers no assurance that enough stress has been relieved to prevent another quake.

<u>Figure 2</u>

<u>Types of Faults</u>

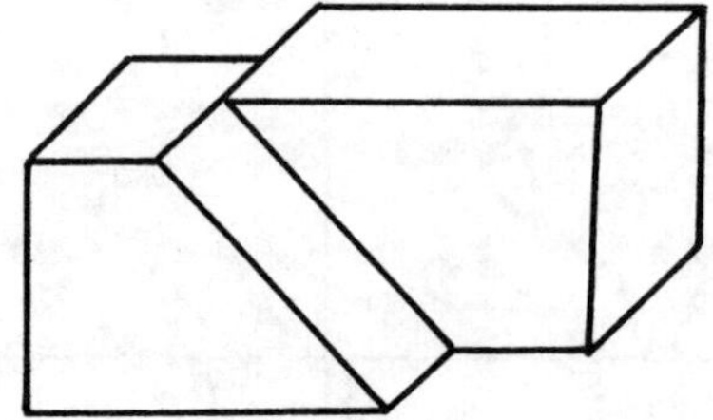

Above: left-lateral strike slip fault
Below: normal dip-slip fault

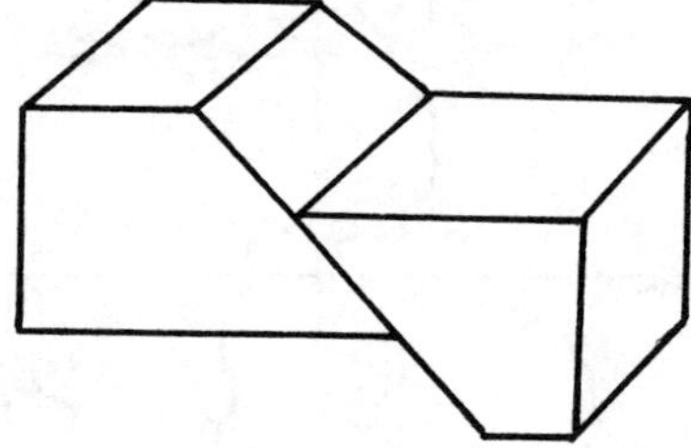

The focal depth of an earthquake is the depth from the Earth's surface to the region (focus) where an earthquake's energy originates. Earthquakes with focal depths less than 60 kilometers are classified as shallow. Earthquakes with focal depths from 60 to 300 kilometers are classified as intermediate. The focus of deep earthquakes may reach depths of 700 kilometers. Earthquakes in California along the San Andreas and associated faults have shallow focal depths. For most, the depth is less than 10 miles.

Very shallow earthquakes are probably caused by fracturing of the brittle rock in the crust or by internal stresses that overcome the frictional resistance locking opposite sides of a fault. The cause of intermediate and deep earthquakes is not yet fully understood.

The epicenter of an earthquake is the point on the Earth's surface directly above the focus. The location of an earthquake is commonly described by the geographic position of its epicenter and by its focal depth.

<u>C. Plate Tectonics</u>

<u>1. Introduction and Background</u>

Most earthquakes are a manifestation of the fragmentation of the Earth's outer shell into large and small plates, all moving relative to all others with steady velocities that reach 13 centimeters per year, pulling apart here, slipping past one another there, sliding beneath another somewhere else, and in other places colliding slowly to build mountain ranges.

We know the new field which studies plate motions as plate tectonics. The "plate" is the basic unit of the system. "Tectonics" (from the Greek word tekton, meaning builder) refers to the processes and products of motions within the Earth.

The reality of continental drift was reasonably well established during the 1930's. The distribution of indicators of past climates required that the continents had moved slowly about the globe. For example, the same 300-million-year age fossilized deposits are found in India and in the Artic.

Reassembly of continents in patterns suggested by their shapes provided remarkable continuity for similar, truncated geologic terrains. A vocal minority of Earth scientists visualized continentals as rafts sliding over the ocean floor. Most geophysicists regarded this drift as impossible.

In the 1950's, the science of paleomagnetism expanded rapidly and provided powerful new evidence for drift. Many rocks, such as volcanic rocks crystallized from lava, contain tiny grains of magnetite and other magnetic minerals which retain the magnetic orientation of the Earth's magnetic field of the time at which the rocks formed. Early studies of this phenomenon had been made during the 1920's. The magnetic orientations of rocks indicated the same fossil latitudes for the various continents as those indicated by paleoclimatology and other geologic and paleontologic criteria.

The 1950's also saw the gathering of an enormous amount of geophysical data from the oceans, particularly by research vessels from American oceanographic institutions. A globe-girdling system of interconnecting submarine ridges was recognized. Such ridges were located approximately midway between continents (like Africa and South America) that appeared to be moving apart.

It was suggested that new oceanic crust was being formed at the ridges and added to great plates moving apart. Great submarine trenches were also delineated, particularly along the convex oceanic sides of the volcanic arcs which make up the "ring of fire" around the Pacific. From these trenches inclined zones of earthquakes dip as deep as 700 kilometers into the mantle beneath and behind the volcanic arcs. It was reasoned that oceanic crust was formed at spreading ridges behind drifting plates. The crust was destroyed at the same rate elsewhere as oceanic plates tipped down at the trenches and slid deep into the mantle along the seismic zones.

English scientists suggested that the alternating belts of highly and weakly magnetic oceanic crust, known by then to trend along the mid-ocean ridges, represented magnetic recordings as new crust formed in the gap behind separating plates. The belts would record symmetrically, on opposite sides of the young ridges, the periods of time represented by the magnetic-chronology time scale. They tested the hypothesis against several small ridge sectors and showed that it was correct.

2. The Concepts

The Earth's crust is broken into moving plates of "lithosphere" (figure 3). There are seven very large plates, each consisting of both oceanic and continental portions and of a dozen or more small plates (not all of which are shown on figure 3). Each plate is about 80 kilometers thick and can be pictured as having a shallow and a deep part. The shallow part deforms by elastic bending or by brittle breaking. The deep part yields plastically. Beneath the plate is a viscous layer on which the entire plate slides. The plates tend to be internally ridgid, interacting mostly at their edges.

Figure 3

Lithosphere Plates of the World

Lithosphere plates of the world showing boundaries that are presently active. Heavy broken line: zone of spreading from which plates are moving apart. Line with barbs: zone of underthrusting (subduction) where one plate is sliding beneath another; barbs on overriding plate. Dotted line: strike-slip fault, along which plates are sliding past one another. Stippled area: part of a continent, exclusive of that along a plate boundary, which is undergoing active extensional, compressional, or strike-slip faulting. Much simplified in complex areas.

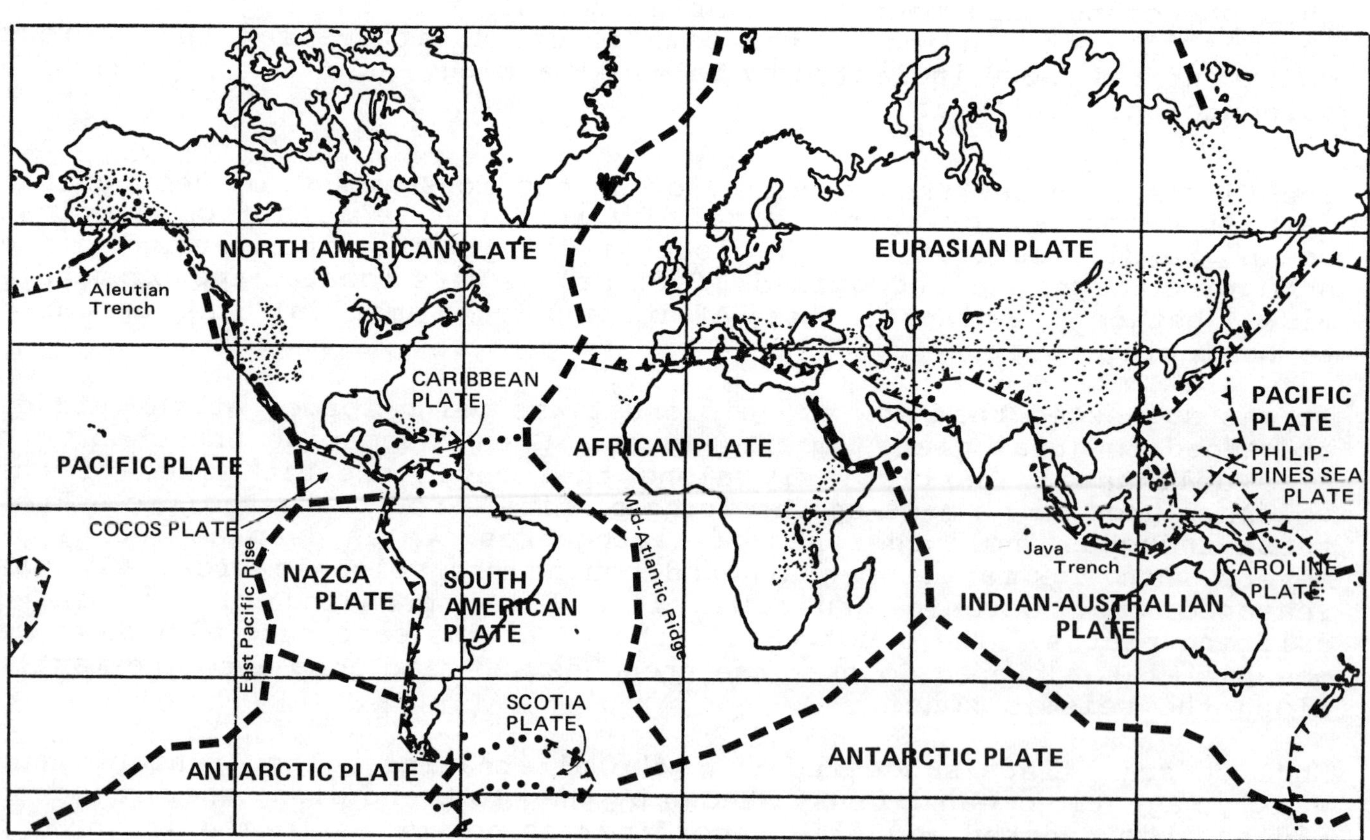

Spherical geometry requires that any motion between two portions of a spherical surface can be expressed as the rotation of one relative to the other, defined by an angular displacement and by a pole of relative rotation. It is also required that all trajectories of relative motions be along small circles to that pole. (Latitude lines are circles relative to the Earth's geographic pole.)

These theoretical constraints are largely met by all of the criteria by which plate motions are demonstrated. The magnetic "stripes" along the midocean ridges define spherical angles. Velocities of separation of adjacent plates are angularly (not linearly) constant. The direction of slip in earthquakes along plate boundaries, and the orientation of strike-slip faults that form or offset those boundaries, are in the corresponding circular directions.

All plates are moving relative to all others. There are grounds for suggesting that the African plate may now be approximately fixed relative to the deep mantle. If so, it is the only such plate. Velocities of relative motion between adjacent plates range from less than 1 centimeter to about 13 centimeters per year.

Although these velocities are slow by human standards, they are extremely rapid by geologic ones. For example, a motion of 5 centimeters per year adds up to 50 kilometers in only 1 million years. Some plate motions have been continuous for 100 million years.

Plates are now pulling apart, primarily along the system of great submarine ridges in the world's oceans. Hot material from the deeper mantle wells up into the gap. Some of it melts and is erupted on the surface as lava or is injected near the surface to crystallize as other igneous rocks. The ridge stands high because its material is hot, and hence, low in density.

As the plates move apart, the ridge material gradually cools and contracts, and its surface sinks. Ridges generally form steplike alternations of spreading centers perpendicular to the direction of motion and of strike-slip faults parallel to that direction.

Where plates converge, one tips down and slides beneath the other. Generally, an oceanic plate slides ("subducts") beneath a continental plate (e.g., along the west coast of South America) or beneath another oceanic plate (e.g., the east side of the Philippine Sea plate). A trench is formed where the undersliding plate tips down. The ocean-floor sediment it carries is scraped off against the front of the overriding plate (figure 4).

We now know much about the mechanics of these junctions from geophysical studies and particularly from seismic-reflection profiles made across them. Farther back under the overriding plate, zones of earthquakes inclined down into the mantle to depths that reach 700 kilometers, show the trajectory of the descending plate.

Typically, a belt of volcanoes lies above this inclined earthquake zone, which is on the average about 125 kilometers deep. The melting which ultimately produces the volcanoes may start with the forcing out of water, previously combined in the crystal structures of various min-

Figure 4

Zone of Subduction

Cross section through a subduction system where an oceanic lithosphere plate is converging with and sliding beneath a continental plate. Details and dimensions are those for western Java and the Java Trench system, but other continental-margin systems are similar.

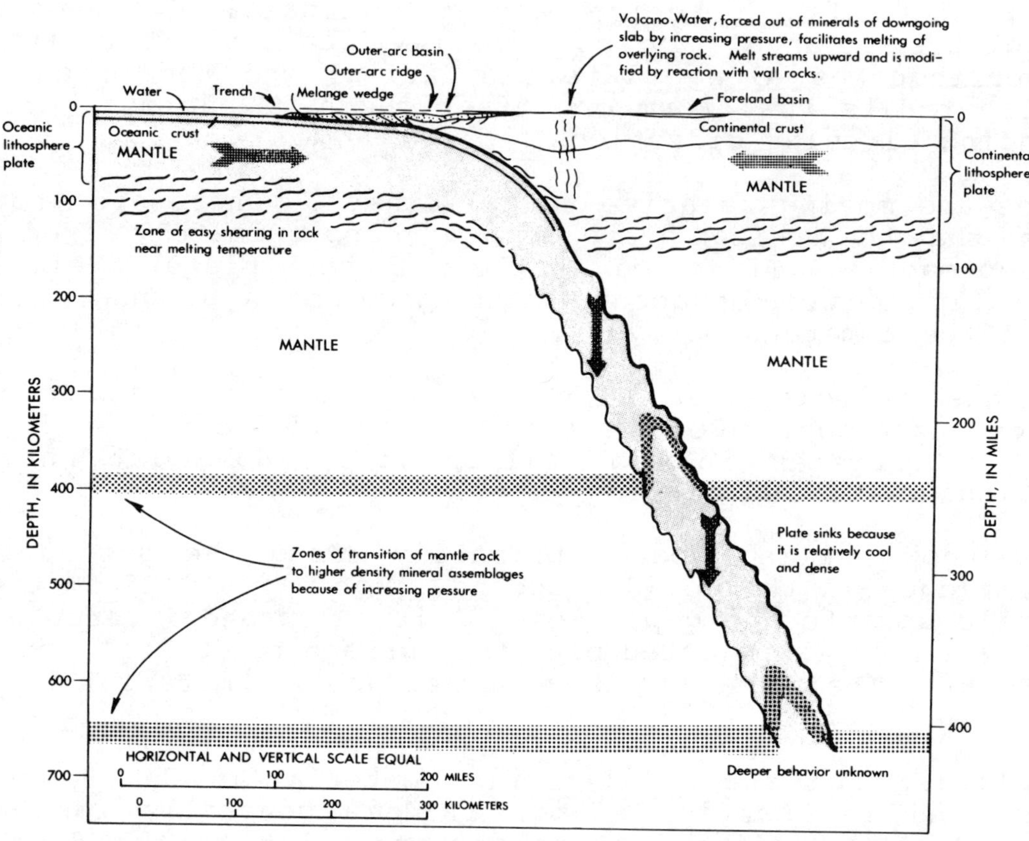

erals, by the increase of pressure on the downgoing plate. The released water lowers the melting temperature, and so permits melting of surrounding rock.

New oceanic-plate (lithosphere) material is generated by the upwelling processes at spreading ridges. Old lithosphere is consumed and recycled deep into the mantle at the same rate as the convergent trenches. The balance is global only. The formation of lithosphere at the Mid-Atlantic Ridge is compensated by subduction primarily in the western Pacific.

3. Earthquake Production

Most earthquakes occur along or near the boundaries of lithosphere plates. Narrow earthquake belts outline the plates. Narrow zones of earthquakes follow the spreading centers and strike-slip faults shown on figure 3. Inclined zones of earthquakes extend deep into the mantle from the indicated subduction-zone trenches. There are also many earthquakes within those continental terrains undergoing distributed deformation.

Shallow earthquakes represent sudden slippages that release strain stored elastically in rock over long periods. It is uncertain whether deep-mantle subduction-zone earthquakes represent similar elastic releases or abrupt contractions of part of a descending lithosphere plate into rock of higher-density.

Only a fraction of the energy released in an earthquake goes into the seismic waves. Most is absorbed locally by moving, deforming, and heating rock. The proportion so absorbed increases irregularly with the increasing size of earthquakes. Minor earthquakes do not generally represent a safety valve adequate to dissipate the energy that is stored for great earthquakes, although slow creep along a fault can provide at least a partial release. On the other hand, the occurrence of a great earthquake does not prove that all dangerous strain has been released.

The magnitude of an earthquake depends upon the length and breadth of the fault plane of slippage, as well as on the amount of slip. The spot location ("epicenter") given for an earthquake is the site above the inital breakage. From such a site, the slip propagates along the fault with a velocity up to that of the outward-radiating seismic shear-wave front (about 3 kilometers per second) until the entire effected segment can be in motion at once.

Oceanic spreading centers are so hot at shallow depths that the solid rock above them cannot store enough elastic strain to produce great earthquakes. The infrequent large earthquakes that do occur in ridge systems are mostly on the longer strike-slip ("transform") faults. Great earthquakes occur primarily along convergent (subducting) and strike-slip plate boundaries and within those parts of the continents undergoing intraplate deformation (figure 1).

4. Seismic Sea Waves

When the sea floor is raised sudddenly during a great earthquake, water is raised with it. If the sea surface is tipped, water rushes away. If the floor is dropped, water rushes in. An enormous mass of water is suddenly set in motion, and complex sloshing back and forth continues for many hours. The result is a train of water waves known as seismic sea waves or "tsunami" (their Japanese name).

The velocity of a wave increases with its length. In the deep ocean, waves travel at about 800 kilometers per hour. The waves at sea may be an hour apart and only perhaps 30 centimeters high. Thus, they are virtually undetectable. As a wave approaches land, the wave velocity decreases due to increased friction with the shallowing sea floor. As the wave velocity decreases, its height increases.

Where sea-floor topography and orientation are optimal for a tsunami, the wave can form a breaking wall of water 15 meters or more high and rush onto the shore to cause enormous destruction. Nearby coastal points, where the bottom configuration is different, may record the same wave only as a rapid surge and withdrawal of water.

The most extensive sudden changes in depth of the sea floor, and hence the greatest seismic sea waves, result from shallow subduction-zone earthquakes, such as the Alaska earthquake of 1964.

D. Active Faults in California

The most earthquake-prone areas in the United States are those that are adjacent to the San Andreas fault system of coastal California and the fault system that separates the Sierra Nevada from the Great Basin. Many of the individual faults of these major systems are known to have been active during the last 150 to 200 years. Others are believed to have been active since the end of the last great ice advance about 10,000 years ago.

Earthquakes in California are relatively shallow and clearly related to movement along active faults. Within the last 150 years, at least 25 California earthquakes have been associated with movements that ruptured the Earth's surface along faults shown on figure 1. These earthquakes are briefly described in table 1. On the San Andreas fault, eight moderate-to-severe earthquakes have been accompanied by movements on the fault at the Earth's surface since 1838. Other faults in the California region have also experienced repeated earthquakes.

The magnitude of shallow earthquakes can generally be correlated with the amount and length of the associated fault movement. Thus, the largest episodes of fault movement (or fault slip) recorded in California accompanied the great earthquakes of 1857, 1872, and 1906--all of which had estimated magnitudes that were over 8 on the Richter Scale.

During the last several hundred years, many of the faults shown on figure 5 have experienced sudden slip and/or creep. For other faults, however, recent activity can only be inferred from geologic and topographic relations which indicate that they have been active during the past several thousand years. Such activity suggests that some of these faults could slip or creep again.

Most of the faults shown on figure 5 are vertical or near vertical breaks. Movement along these breaks has been predominately horizontal. The direction of ground movement along the faults is indicated on the map by arrows. If the block on the opposite side of the fault from the observer has moved to the right, the movement is termed right-lateral. The fault is said to be a strike-slip or wrench fault. Movement of the opposite block to the left is termed left-lateral. Note that most of the faults trend northwesterly and that the movement has been right-lateral. Notable exceptions to the predominant north-westerly trend of faults are the west-trending Garlock and Big Pine faults. Movement on these faults has been left-lateral.

A few reverse faults have also been active in California. The planes of such faults are inclined to the Earth's surface. The rocks above the fault plane have been thrust upward over the rocks below the fault plane. The magnitude 7.7 Arvin-Tehachapi earthquake of 1952 (number 20 in table 1) was associated with such movement along the White Wolf reverse fault. The magnitude 6.6 San Fernando earthquake of 1971 (number 25 in the table) was caused by a sudden rupture along a reverse fault at the foot of the San Gabriel Mountains.

Figure 5

Active Faults in California

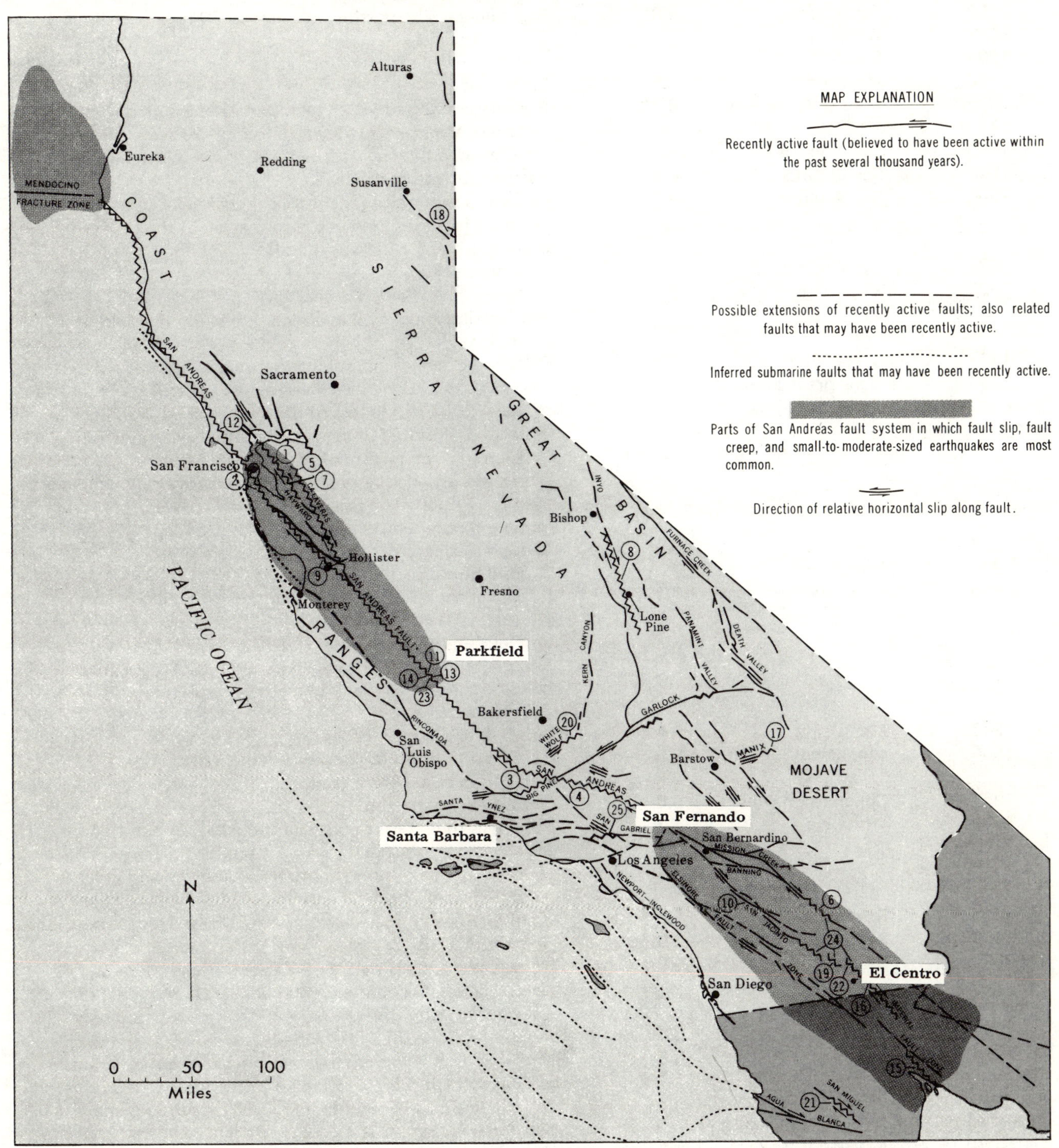

Table 1

California Earthquake Chronology

DATE	FAULT	RICHTER MAGNITUDE	SURFACE EFFECTS AND SIGNIFICANCE
1836	Hayward	7.0*	Ground breakage
1838	San Andreas	7.0*	Ground breakage
1852	Big Pine		Possible ground breakage
1857	San Andreas	8.0*	Right-lateral slip, possibly as much as 30 feet
1861	Calaveras		Ground breakage
1868	San Andreas		Long fissure in earth at Dos Palmas
1868	Hayward	7.0*	Strike-slip
1872	Owens Valley zone	8.3*	Right-lateral slip of 16-20 feet. Left-lateral slip may also have occurred. Vertical slip, down to east, of 23 feet.
1890	San Andreas		Fissures in fault zone; railroad tracks and bridge displaced.
1899	San Jacinto	6.6*	Possible surface evidence
1906	San Andreas	8.3	Known as the "San Francisco earthquake." Right-lateral slip up to 21 feet. Resulted in the formation of the California State Earthquake Investigation Commission.
1922	San Andreas	6.5	Ground breakage
1925	Mesa/Santa Ynez	6.3	Known as the "Santa Barbara earthquake of 1925." U.S. Coast and Geodetic Society was directed to study the field of seismology.
1927	Santa Ynez	7.5	Occurred offshore in a submarine trench, but was still felt.
1933	Newport-Inglewood	6.3	Known as the "Long Beach earthquake of 1933." Extensive property damage and loss of life. Many school buildings were destroyed. Resulted in the passage of the Field Act. The Division of Architecture of the State Department of Public Works was assigned responsibility to approve new buildings used for schools. The Riley Act was also passed, and it set minimum requirements for lateral force design.
1934	San Andreas	6.0	Ground breakage
1934	San Jacinto	7.1	Ground breakage
1940	Imperial	7.1	Known as the "El Centro earthquake." 40 miles of surface faulting. 80% of Imperial buildings were damaged. However, no Field Act school buildings were damaged. This was the first major earthquake to yield accelerograph data on building periods. A maximum acceleration (ground) of .33g was experienced.
1947	Manix	6.4	Left-lateral slip of 3 inches
1950	(unnamed)	5.6	Vertical slip, down to west, of 5-8 inches along the west edge of Fort Sage Mountains
1951	Superstition Hills	5.6	Slight right-lateral slip
1952	White Wolf	7.7	Known as the "Kern County earthquake," and the "Arvin-Tehachapi earthquake." Extensive building damage to old buildings. Little or none to properly designed and Field Act buildings. Confirmed the requirement for proper design.
1956	San Miguel	6.8	Right-lateral slip, 3 feet; vertical slip, down to southwest, 3 feet.
1966	Imperial	3.6	Right-lateral slip, ½ inch.
1966	San Andreas	5.5	Known as the "Parkfield earthquake." Right-lateral slip of several inches. Maximum ground acceleration of .50g - highest recorded to date.
1968	Coyote Creek	6.4	Right-lateral slip up to 15 inches
1971	San Fernando	6.6	Known as the "San Fernando earthquake." Left-lateral slip up to 5 feet; north-side thrusting up 3 feet. Massive instrumentation due to 1965 Los Angeles building code resulted in more than 300 accelerograph plots. 1.24g experienced at Pacoima dam.
1979	Imperial	6.6	Known as the "Imperial Valley earthquake." Right-lateral slip up to 55 cm with more than 30 km of surface rupture. Extensive accelerograph data collected. Resulted in the first accelerograph data from an extensively damaged building (Imperial County Services Building.)

*estimated

Figure 6

Right and Left-Lateral Faults

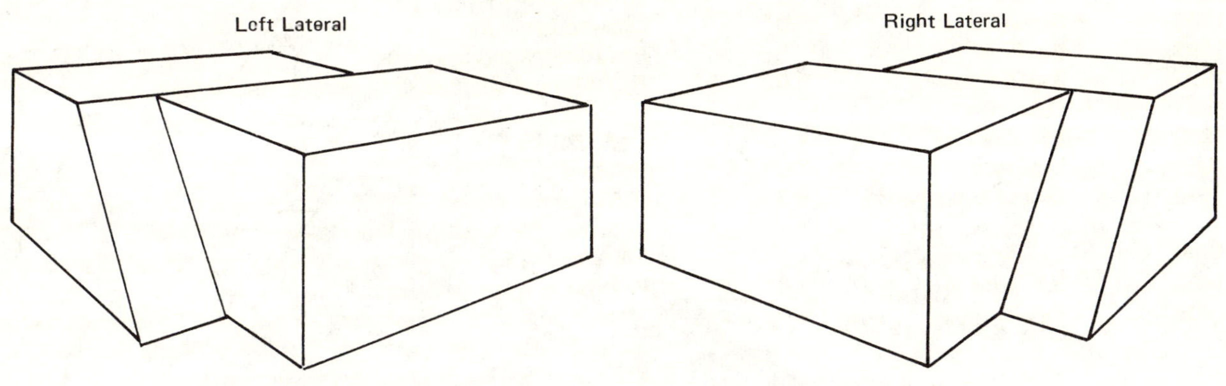

Studies of fault movements have shown that they occur in two ways. The first and better understood is a sudden displacement (slip) of the ground along a fault. Such displacement is accompanied by earthquakes and occasionally produces offsets of topographic and man-made features. During the 1906 earthquake, the ground was displaced as much as 21 feet along the San Andreas fault in northern California. During the 1857 earthquake, displacement of the ground along this fault was as much as 30 feet in southern California.

Figure 7

Evidence of Fault Slip

Fence offset 8½ feet near Woodville, California, as a result
of the historic San Francisco earthquake, April 18, 1906.

The second type of fault movement, known as creep, is now taking place on portions of faults in California. Creep is characterized by continuous or intermittent movement without noticeable earthquakes. Recent fault creep on portions of the Hayward, Calaveras, and San Andreas faults has produced cumulative offsets ranging from a fraction of an inch to almost a foot in curbs, street, and railroad tracks and has caused some damage to buildings.

Figure 8

Major Faults of the San Francisco Bay Area

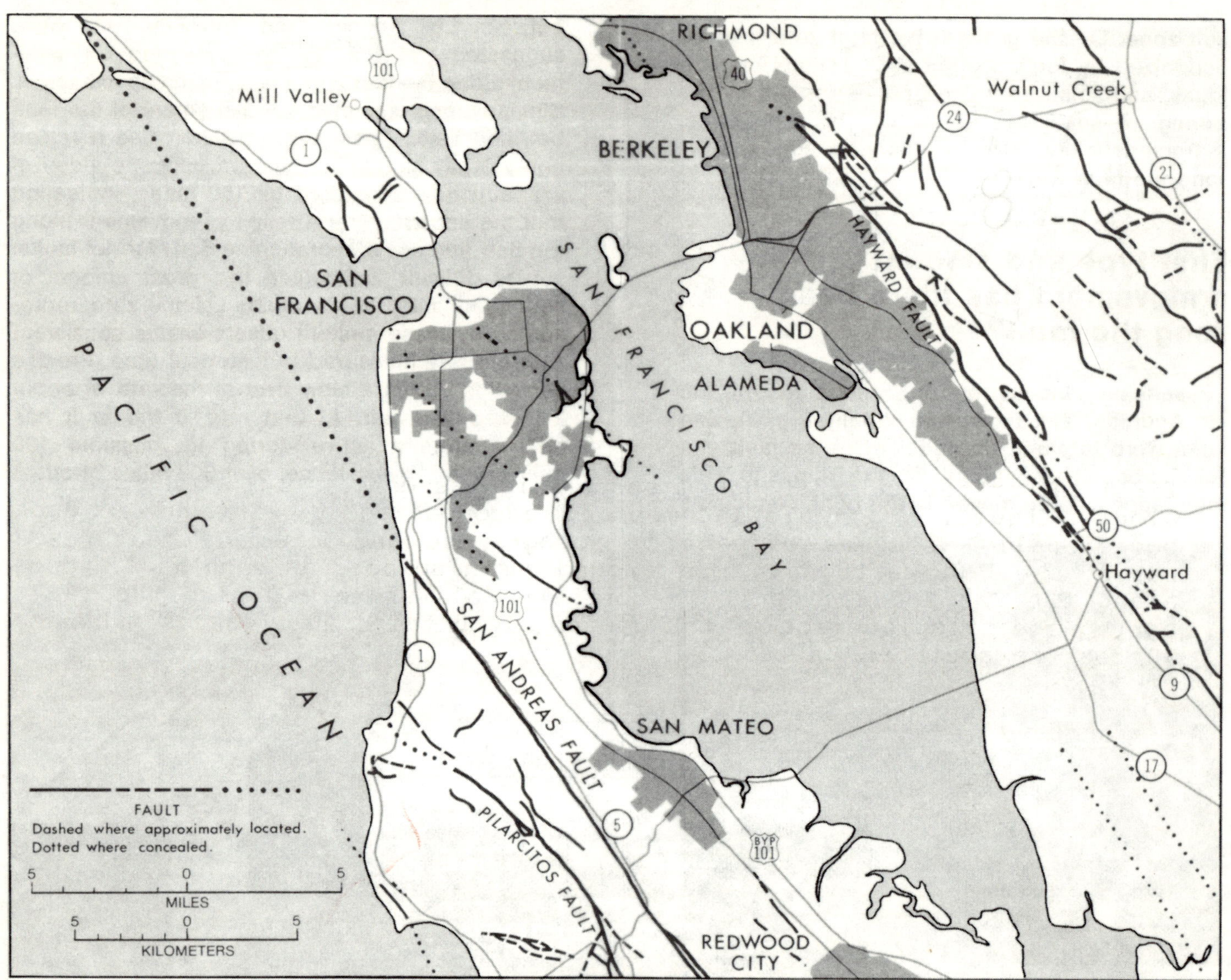

While difficult to imagine this great amount of shifting of the Earth's crust, the distance represented by these offsets seems consistent with the rate measured. Precise surveying shows a slow drift at the rate of about 2 inches per year. At that rate, over 350 miles of offset is a possibility if the fault has been uniformly active during its possible 100 million years of existence.

Figure 10 shows the location of some of the larger earthquakes in the California-Nevada region. The 1838, 1857, and 1906 earthquakes are the largest that have occurred along the San Andreas fault.

Most of the faults shown in figure 5 are zones made up of a number of subsidiary faults or fault strands. These fault zones range in width from several feet to a mile or more. Slip along them has been found to recur repeatedly on only one or a few of the multiple strands that constitute these zones. Most of the strands show no evidence of recent activity. However, slip does at times recur on older trands or on entirely new ones. The strong tendency for fault slip and earthquakes to recur along the most recently active strands makes knowledge of the precise location of these strands essential to land-use planning.

Since 1934, earthquake activity along the San Andreas fault system has been concentrated in three areas shown by shading on the map. These are areas where earthquakes and fault displacements of the Earth's surface have been common and where fault creep is currently taking place. The zones between these three active areas have had almost no earthquakes or known slip events since the great earthquakes of 1857 in the southern segment and 1906 in the northern segment. This implies that these two zones of the San Andreas fault are temporarily locked. In the other areas, stress is being continually relieved by creep. The lack of such activity in the locked segments could mean that these segments are subject to less frequent but larger fault movements and correspondingly more severe earthquakes.

E. The San Andreas Fault

The San Andreas fault is the master fault of a fault network that cuts through rocks of the California coastal region. This right-lateral fault is a huge fracture 600 or more miles long. It extends almost vertically into the Earth to a depth of at least 20 miles. In detail, it is a complex zone of crushed and broken rock from a few hundred feet to a mile wide. Many smaller faults branch from and join the San Andreas fault zone.

Figure 5 shows the general location of the San Andreas and other major faults in California. Figure 8 gives greater detail.

Over much of its length, a linear trough reveals the presence of the fault. From an airplane, the linear arrangements of the lakes, bays, and valleys appear striking. However, people may not realize they are on the San Andreas fault zone as they drive near Crystal Springs Reservoir, along Tomales Bay, or through the Cajon or Tejon pass. On the ground, the fault zone can be recognized by long straight escarpments, narrow ridges, and small undrained ponds formed by the settling of small blocks within the fault zone. Characteristically, stream channels jog sharply along the fault trace.

Geologists who have studied the fault between Los Angeles and San Francisco have suggested that the total accumulated displacement along the fault may be as much as 350 miles. Similarly, geologic study of a segment of the fault between Tejon Pass and the Salton Sea has revealed geologically similar terrains on opposite sides of the fault now separated by 150 miles. This indicates that the separation is a result of movement along the San Andreas and branching San Gabriel faults.

Figure 9

A Portion of the San Andreas Fault

Figure 10

Epicenters of California Earthquakes
(Heavy lines show ground breakage.)

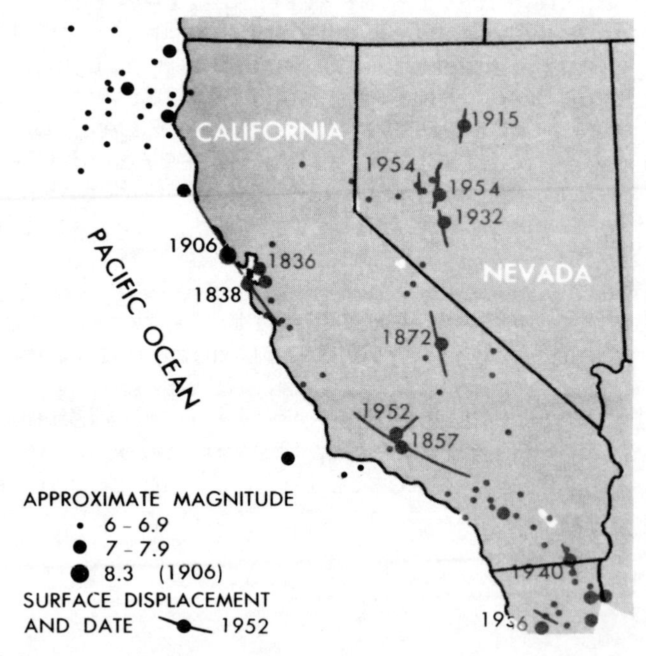

5. Intensity and Magnitude

The severity of an earthquake can be expressed in terms of both intensity and magnitude. However, the terms are different and are often confused.

Intensity is based on the observed effects of ground shaking on people, buildings, and natural features. It varies from place to place within the disturbed region, depending on the location of the observer with respect to the earthquake epicenter.

Magnitude is related to the amount of seismic energy released at the hypocenter of the earthquake. It is based on the amplitude of the earthquake waves recorded on instruments which have a common calibration. The magnitude of an earthquake is thus represented by a single instrument reading.

A. The Richter Magnitude Scale

Seismic waves are vibrations that travel through the Earth. They are recorded on seismographs. Seismographs record the varying amplitude of ground oscillations beneath the instrument. Sensitive seismographs greatly magnify these ground motions and can detect strong earthquakes from sources anywhere in the world. The time, location, and magnitude of an earthquake can be determined from the data recorded by seismograph stations.

In 1935, Charles F. Richter of the California Institute of Technology developed the Richter magnitude scale to compare the size of earthquakes. The magnitude of an earthquake is determined from the logarithm of the amplitudes recorded by seismographs. Adjustments are included in the magnitude to compensate for the variation in the distance between the various seismographs and the epicenter of the earthquakes.

On the Richter Scale, magnitude is expressed in whole numbers and decimal fractions. For example, a magnitude of 5.3 might be computed for a moderate earthquake. A strong earthquake might be rated as magnitude 7.3. Because of the logarithmic basis of the scale, each whole number increase in magnitude represents a tenfold increase in measured amplitude. Earthquakes with magnitudes of about 2.0 or less are usually called microearthquakes. They are not commonly felt by people and are generally recorded only on local seismographs.

Events with magnitudes of about 4.5 or greater (there are several thousand such shocks annually) are strong enough to be recorded by sensitive seismographs all over the world. Earthquakes with magnitudes below 4.5 have little potential to cause structural damage. Great earthquakes, such as the 1906 San Francisco earthquake and the 1964 Alaskan earthquake, have magnitudes of 8.0 or higher. On the average, one earthquake of such size occurs somewhere in the world each year. Although the Richter Scale has no upper limit, the largest known shocks have had magnitudes in the 8.8 to 8.9 range.

The Richter magnitude of an earthquake which produces a maximum amplitude (A) in millimeters is

$$M = \log_{10}(A) - \log_{10}(A_o) = \log_{10}(A/A_o) \qquad 1$$

A_o is the seismograph reading produced by an earthquake of standard size (e.g., calibration earthquake). Generally, A_o is .001 of a millimeter. Equation 1 assumes a distance of 100 km separates the seismograph and the epicenter. For other distances, the nomograph of figure 11 may be used to calculate the magnitude.

Assuming that the frequency distribution of earthquakes in California is the same as for the rest of the world, the number of earthquakes of any given magnitude can be determined from table 2. This table gives the number of earthquakes per 100 years for the entire state covering 150,000 square miles. It does not give the frequency of occurrence for any particular location.

Table 2

Frequency of Occurrence

Magnitude	Number per 100 Years
4.75 – 5.25	250
5.25 – 5.75	140
5.75 – 6.25	78
6.25 – 6.75	40
6.75 – 7.25	19
7.25 – 7.75	7.6
7.75 – 8.25	2.1
8.25 – 8.75	.6

Once the Richter magnitude is known, an approximate relationship (equation 2) can be used to calculate the energy radiated. The energy radiated is less than that released by the earthquake due to the heat generation and other non-elastic effects which are not included in equation 2. Little is known about the percentage of the release energy that is actually radiated.

$$\log_{10}[e_{radiated}] = 11.8 + 1.5(M) \quad \text{(in ergs)} \qquad 2$$

The fault length L (in kilometers) and the vertical or horizontal offset D (in centimeters) can also be related to the Richter magnitude.

$$\log_{10}(L) = 1.02(M) - 5.77 \qquad 3$$

$$\log_{10}[10,000)(L)(D^2)] = 1.90(M) - 2.65 \qquad 4$$

B. Seismic Waves

Seismic waves are of two types: compression and shear. Both types of waves pass through the Earth's interior from the focus of an earthquake

Figure 11
RICHTER SCALE NOMOGRAPH

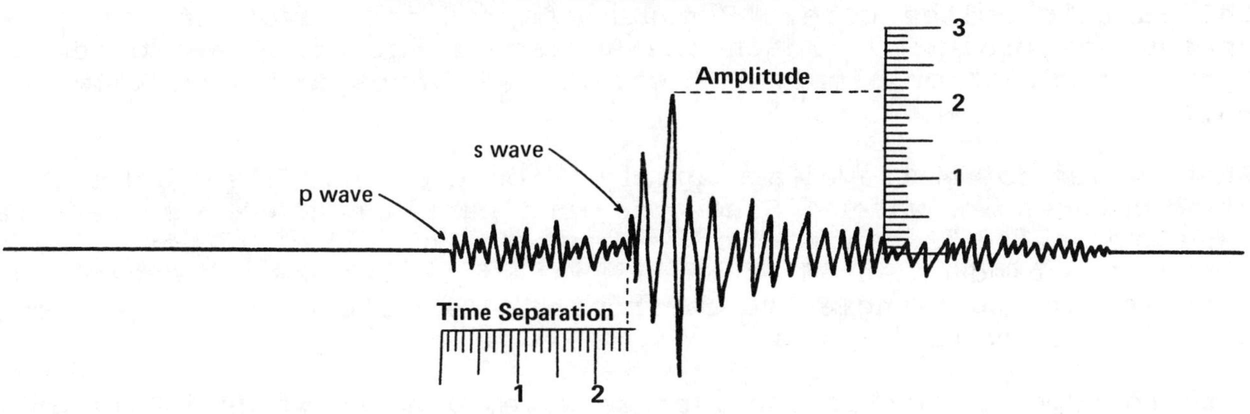

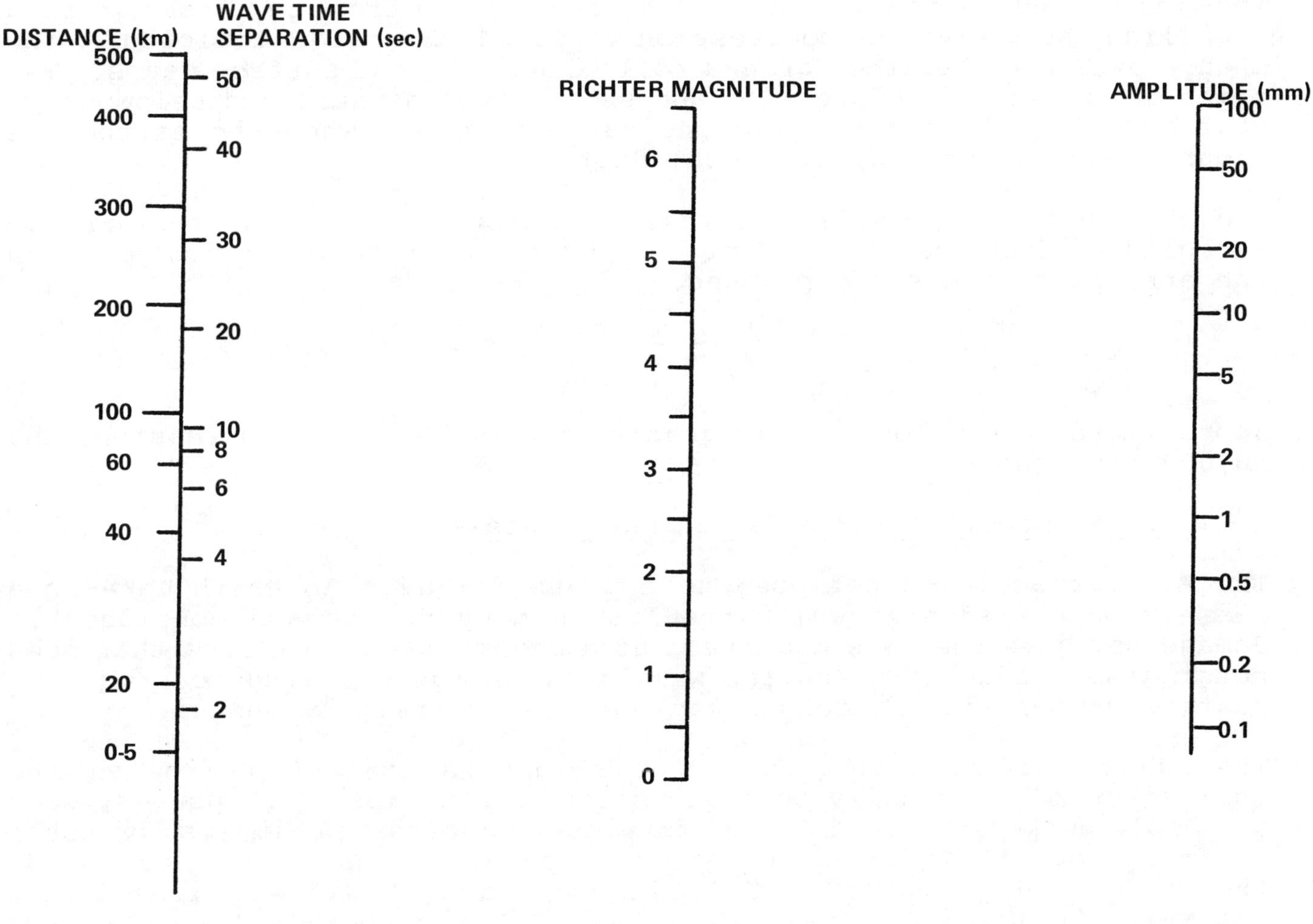

step 1: Determine the time separation between the arrival of the p and s waves. In the above seismogram tracing, the time separation is approximately 24 seconds.

step 2: Determine the maximum amplitude of oscillation. In the above seismogram tracing, the amplitude is approximately 22 millimeters.

step 3: Connect the time separation (on the left scale) and the amplitude (on the right scale) with a straight line. Read the Richter magnitude from the center scale.

step 4: Read the distance from the seismograph to the epicenter from the distance scale.

to distant points on the surface. Only compression waves travel through the Earth's molten core. Because compression waves travel at great speeds and ordinarily reach the surface first, they are often called "primary waves" or simply "p" waves. (P waves are also known as "c" waves.)

Shear waves do not travel as rapidly through the Earth's crust and mantle as do compression waves. Since they ordinarily reach the surface later, they are called "secondary" or "s" waves. Instead of affecting material directly behind or ahead of their line of travel, shear waves displace material at right angles to their path. Therefore, they are sometimes called "transverse" waves.

A third type of wave is the surface wave, also known as a Raleigh ("R") wave or Love ("L") wave. Such waves may or may not form. Their arrival is after the compression and shear waves.

The first indication of an earthquake will often be a sharp thud signaling the arrival of compression waves. This will be followed by the shear waves and then the "ground roll" caused by the surface waves. The times of arrival of compression and shear waves at selected seismograph stations throughout the world indicate where and when the earthquake occurred and sometimes, its focal depth.

The distance, d, from a recording station to an epicenter can be established from the wave velocities and the difference in time between the arrival of the s and p waves.

$$(t_s - t_p) = (\frac{1}{v_s} - \frac{1}{v_c})d \qquad\qquad 5$$

Approximate wave velocities in granite are 19,000 fps (compression) and 10,000 fps (shear).

C. The Modified Mercalli Intensity Scale

The Richter Scale is not used to express damage. An earthquake in a densely populated area which results in many deaths and considerable damage may have the same magnitude as a shock in a remote area that does nothing more than frighten the wildlife. Large magnitude earthquakes that occur beneath the oceans may not even be felt by humans.

The effect of an earthquake on the Earth's surface is called the intensity. The intensity scale consists of a series of responses, such as people awakening, movement of furniture, damage to chimneys, or total destruction. Although numerous intensity scales have been developed, the one currently used in the United States is the Modified Mercalli (MM) Intensity Scale.

The Modified Mercalli Scale was developed in 1931 by the American seismologists Harry Wood and Frank Neumann. The scale is composed of 12 increasing levels of intensity that range from imperceptible shaking to catastrophic destruction. Intensity is designated by Roman numerals. The scale does not have a mathematical basis. Instead, it is an arbitrary ranking based on observed effects.

The Modified Mercalli Intensity value is more meaningful to the non-scientist than the Richter magnitude because intensity refers to the effects actually experienced. After the occurrence of widely-felt earthquakes, the U.S. Geological Survey mails questionnaires to the disturbed area requesting information so intensity values can be assigned. The results are used to compile isoseismal maps that show the extent of various levels of intensity within the felt area.

The lower numbers of the intensity scale generally deal with the manner in wh.ch the earthquake is felt by the people. The higher numbers of the scale are based on observed structural damage. Structural engineers usually contribute information for assigning intensity values of VIII or above.

Table 3 is an abbreviated description of the 12 levels of intensity.

6. Ground Motion Measurements

There are many methods of detecting and measuring earthquakes, although the seismograph and accelerograph are most widely used.

A. Seismograph and Seismometer

A seismograph is a ground-mounted device which measures the actual displacement of the ground with respect to a stationary reference point. The earth movement is recorded by an optical or pen tracing known as a seismogram. A time scale is also recorded on the seismogram to accurately determine the arrival times of the waves. A seismograph usually only records motion in one orthogonal direction. Three seismographs are needed to record all components of ground motion.

Because a seismograph is a spring-mass-dashpot device, it will distort (magnify) earthquakes with periods of frequencies in certain ranges. The ratio of actual damping to critical damping can be modified to minimize this distortion. Good seismograph design calls for a damping ratio of between .6 and .7 with a natural period of vibration smaller than the smallest period to be measured. These design principles are incorporated into the Wood-Anderson seismograph which has a damping ratio of .7 and a period vibration of 1.25 hertz.

B. Accelerograph, Accelerometer, and Strong Motion Seismograph

Strictly speaking, an accelerograph is a seismograph. However, the term "accelerograph" is primarily used to describe seismographs mounted in buildings at the site of earthquakes for the purpose of recording large swaying motions only. Such large swings would exceed the scale limits of most seismographs. An accelerograph located in a building does not run continually. It is triggered by the p-wave and runs for a fixed period of time.

C. Tiltmeter

A tiltmeter installed in the earth works on the same principle as a carpenter's level. A bubble floats in a spherical dome. The slightest movement of the bubble is electronically detected to reveal tilting of the earth.

Table 3

The Modified Mercalli Scale

I   Not felt except by very few under especially favorable conditions.

II   Felt only by a few persons at rest, especially by those on upper floors of buildings. Delicately suspended objects may swing.

III   Felt quite noticeably by persons indoors, especially in upper floors of buildings. Many people do not recognize it as an earthquake. Standing vehicles may rock slightly. Vibrations similar to the passing of a truck. Duration estimated.

IV   During the day, felt indoors by many, outdoors by a few. At night, some awakened. Dishes, windows, doors disturbed; walls make cracking sound. Sensation like heavy truck striking building. Standing vehicles rocked noticeably.

V   Felt by nearly everyone; many awakened. Some dishes, windows broken. Unstable objects overturned. Pendulum clocks may stop.

VI   Felt by all, many frightened. Some heavy furniture moved. A few instances of fallen plaster. Damage slight.

VII   Damage negligible in buildings of good design and construction; slight to moderate in well-built ordinary structures; considerable damage in poorly-built structures. Some chimneys broken.

VIII   Damage slight in specially-designed structures; considerable damage in ordinary substantial buildings with partial collapse. Damage great in poorly-built structures. Fall of chimneys, factory stacks, columns, monuments, walls. Heavy furniture overturned.

IX   Damage considerable in specially-designed structures; well-designed frame structures thrown out of plumb. Damage great in substantial buildings, with partial collapse. Buildings shifted off foundations.

X   Some well-built wooden structures destroyed; most masonry and frame structures with foundations destroyed. Rails bent.

XI   Few, if any masonry structures remain standing. Bridges destroyed. Rails bent greatly.

XII   Damage total. Lines of sight and level are distorted. Objects thrown into air.

D. Magnetometer

The deformation of rock under pressure can be measured by a magnetometer. Such deformation changes the magnetic permeability of rock, causing a local change in the magnetic field of the earth.

E. Scintillation Counter

Scintillation counters are installed in wells to measure the amount of radioactive radon gas in the water. Minute amounts of this gas are released into well-water by rocks under stress.

F. Resistivity Gauge

Changes in the resistivity of rock are indications of density and water content changes. Both density and water content can be expected to change during stress (pressure) fluctuation.

G. Creepmeter

A creepmeter is a wire stretched across a fault. Movement of the fault increases the tension in the wire.

H. Gravimeter

A gravimeter responds to variations in the local force of gravity. Such variations are the result of underground changes in the rock density.

I. Laser Ranging Instruments

A laser can be used to measure the round-trip travel time of a light pulse which is directly related to the distance between two points. Such a test can be performed to detect any movement that has occurred between two points.

8. Ground Motion Characteristics

Although there is not an abundance of seismological data on major seismic events, there is sufficient information on small and intermediate events to generalize about earthquake characteristics.

A. Maximum Ground Acceleration

Maximum ground acceleration decreases with the distance from the earthquake fault. However, this decrease is small in the general area of the fault. Table 4 lists maximum ground acceleration versus magnitude. In general, the accelerations are on the high side.

B. Duration of Ground Shaking

Table 4 also lists the expected duration of the strong ground motion. At a distance somewhat removed from the fault, the duration will be longer but the intensity of shaking will be lower.

An attenuated shaking can also be expected for a longer period than is listed in table 4.

Table 4

**Maximum Ground Accelerations and
Durations of Strong Phase of Shaking**

Magnitude	Maximum Acceleration	Duration
5.0	.09 g	2 sec
5.5	.15	6
6.0	.22	12
6.5	.29	18
7.0	.37	24
7.5	.45	30
8.0	.50	34
8.5	.50	37

C. Estimating the Design Magnitude

In order to design a building to meet seismic requirements, it is
necessary to determine the maximum expected earthquake magnitude. There
are two practical methods of doing this.

The first method assumes different shapes of the magnitude versus
frequency of occurrence distributions for different geographical regions. This method is based on a seismic probability map such as is shown
in figure 12. The maximum intensity of shaking in the various zones and
the approximate responding magnitudes are shown in table 5.

Table 5

Maximum Zonal Accelerations

Zone	Maximum Acceleration	Maximum Magnitude
0	.04 g	4.25
1	.08	4.75
2	.16	5.75
3 (not near a great fault)	.33	7.0
4 (near a great fault)	.50	8.5

The second method assumes all regions have magnitude-frequency distributions with the same shape and that earthquakes of all magnitudes are
possible in any region of the world. The frequency distribution giving
the probability of an earthquake of magnitude M or greater is

$$P[>M] = e^{-M/B}$$

6

B is a seismic parameter which has been determined to be .48 for 100,000
square miles of Southern California. (B is 2.1 when all 150,000 square
miles of California are considered.) Equation 6 does not recognize any
upper bound on M. As a practical matter, the probability of M exceeding
8.5 must be considered negligible.

Figure 12

Seismic Zone Map of United States

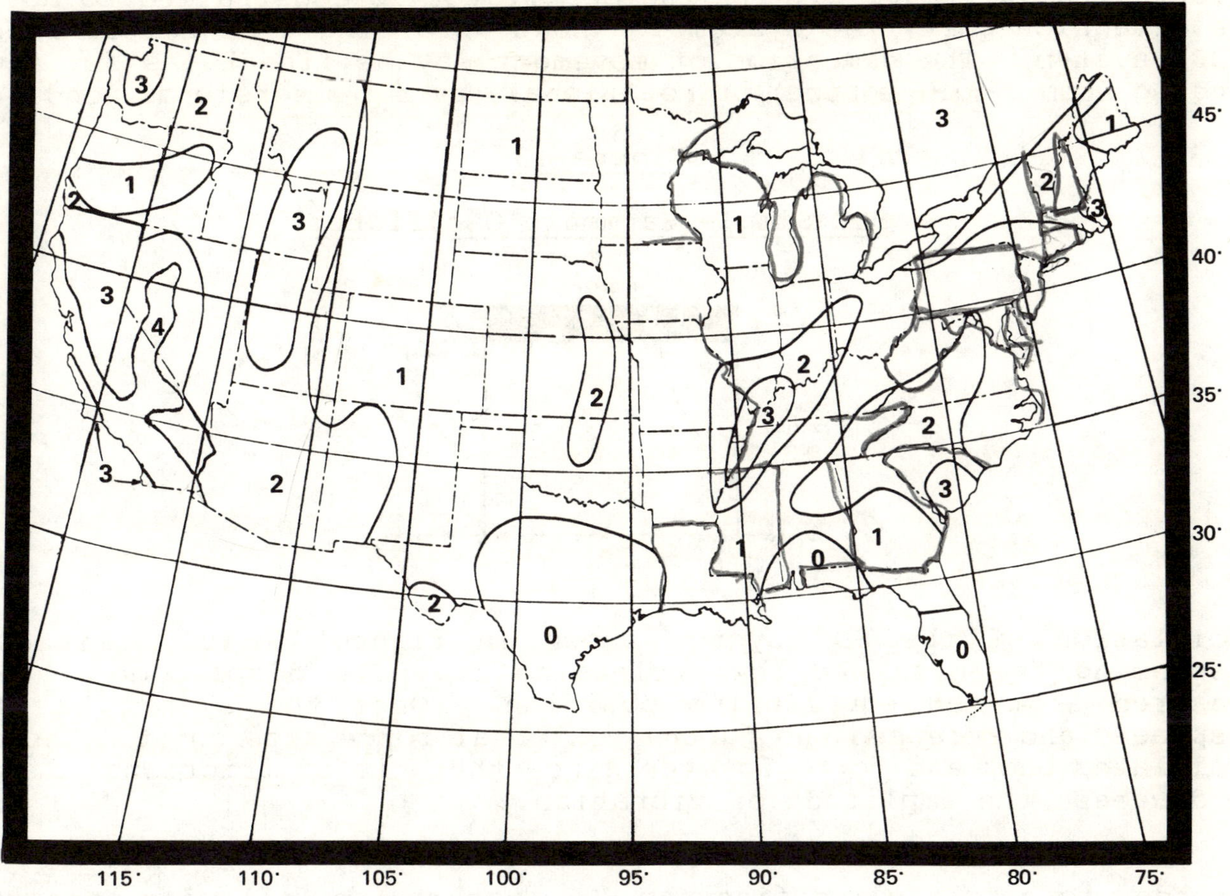

The expected number of earthquakes (E) having magnitudes greater than M during Y years is given by equation 7.

$$E = YN_o e^{-M/B}$$

$$7$$

N_O is a parameter having a value of 1.7 (1/mile^2) for 100,000 square miles of Southern California.

9. Vibration Analysis

There is no need to actually develop the differential equations of vibratory motion for a building structure. However, this section is used to introduce some of the terms and concepts related to structural dynamics.

A. Simple Harmonic Motion

Ideal systems which consist only of springs and masses and which are not acted upon by external disturbing forces (after an initial displacement) are known as simple harmonic oscillators. These oscillators move in a repetitive pattern known as simple harmonic motion. Simple harmonic motion is characterized by the lack of frictional damping. Thus, once set in motion, simple harmonic oscillators remain in motion.

Such a simple system may be visualized as a mass hanging on a spring, a helical coil, a pendulum on a frictionless pivot, or a slab supported by two frictionless and massless cantilever springs. This latter case is illustrated in figure 13. If the oscillator is constrained to move in one dimension only, the system is known as a single degree of freedom (SDF) system. The dimension of movement may be linear (as for a mass hanging from a coil spring) or rotational (as for a torsional pendulum).

<u>Figure 13</u>

<u>SDF Simple Harmonic Oscillator</u>

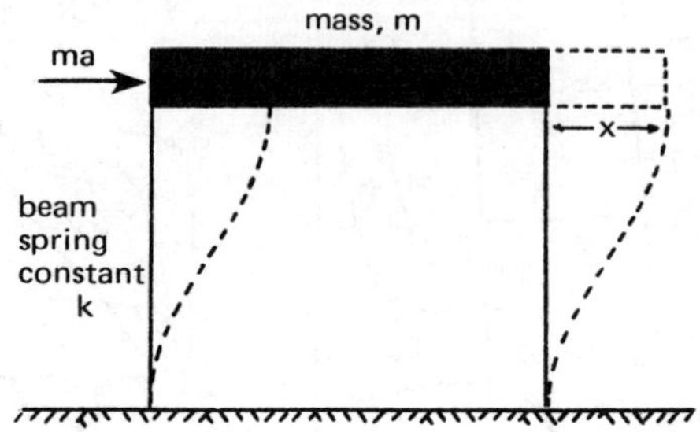

Oscillation of the SDF system shown in figure 13 is initiated by displacing the mass and then releasing it. The displacement (x) is measured from the equilibrium position. Once the system has been displaced and released, no further external force acts on it. The mass oscillates back and forth forever since there is no frictional damping to decrease the amplitude of vibration.

In order to completely define the SDF system, the following parameters are necessary:

    m   the mass of the object

    k   the combined spring constant of the cantilever springs

    x_o the displacement just prior to release

    v_o the velocity just prior to release

After being released, the only forces acting on the mass are the restoring spring force and the inertial restraining force. Therefore, the equilibrium equation of the mass is

$$F_{inertial} + F_{spring} = 0 \qquad\qquad 8$$

However, the spring force is (kx) and the inertial force is given by Newton's second law (ma). Therefore,

$$ma + kx = 0 \qquad\qquad 9$$

Dividing by k and recognizing that acceleration is the second derivative of position results in the following relationship:

$$x'' + (k/m)x = 0 \qquad\qquad 10$$

Since the ratio (k/m) is just a constant, equation 10 is a first order linear differential equation. The solution gives the position, x, as a function of time.

$$x = A(\sin\omega t) + B(\cos\omega t) \qquad\qquad 11$$

A and B are constants given by the formulas below.

$$A = v_o/\omega \qquad\qquad 12$$

$$B = x_o \qquad\qquad 13$$

ω is the circular (angular) frequency of the system given by equation 14. Since no external force acts on the mass after its release, ω is known as the natural frequency of the system.

$$\omega = \sqrt{(k/m)} \qquad\qquad 14$$

The angular frequency, linear frequency, and period of oscillation are related by equation 15.

$$\omega = 2\pi f = 2\pi/T \qquad\qquad 15$$

The maximum value of x is given by equation 16. Notice that the amplitude is equal to x_o if the initial velocity is zero.

$$\text{amplitude} = \sqrt{(v_o/\omega)^2 + (x_o)^2} \qquad\qquad 16$$

The position of the mass as a function of time is shown in figure 14.

Figure 14

Simple Harmonic Motion

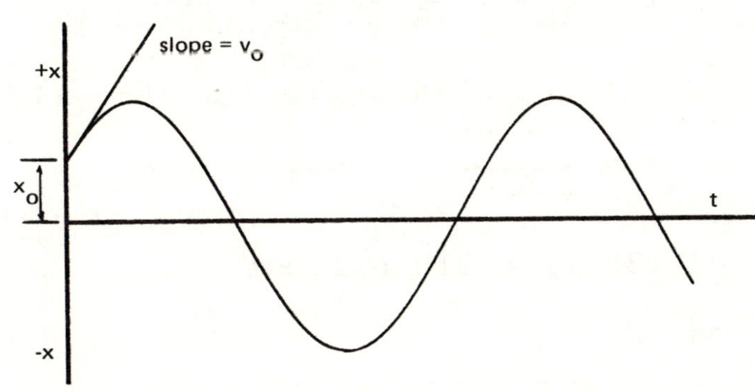

Example 1

What is the natural period of vibration of the system shown below?

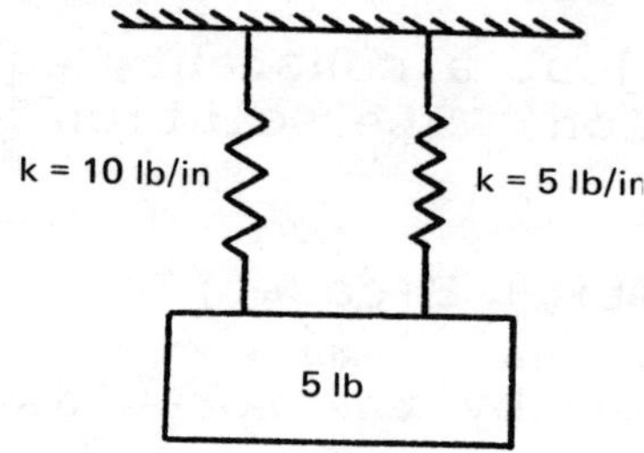

Solution

The composite spring constant is 5 + 10 = 15 lb/in = 180 lb/ft

The object mass is (5/32.2) = .155 slugs

The period is T = $2\pi\sqrt{.155/180}$ = .184 seconds

Example 2

What is the natural period of vibration of the inverted pendulum system shown below? Neglect the support weight.

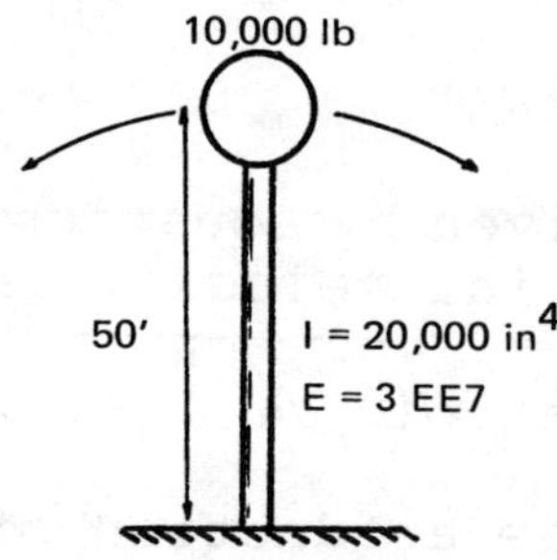

Solution

Consider the pendulum as a cantilever beam. The lateral deflection when a force F of one pound is applied is

$$(1/k) = \frac{L^3}{3EI} = \frac{[(50)(12)]^3}{(3)(3\ EE7)(20,000)}$$

$$= 1.2\ EE\text{-}4\ in/lb = 1\ EE\text{-}5\ ft/lb$$

Then,
$$k = \frac{1}{1\ EE\text{-}5} = 1\ EE5$$

The mass is (10,000/32.2) = 310.6 slugs.

The natural period is

$$T = 2\pi\sqrt{310.6/(1\ EE5)} = .35\ seconds$$

B. Free Systems with Damping

All real systems have friction. A free system with friction will
oscillate with a continuously-decreasing amplitude due to the energy
loss during each swing. The frictional force in an ideal system is
proportional to the velocity of the mass. That is,

$$F_{damping} = Bv \qquad\qquad 17$$

Figure 15 illustrates a free system with damping. The frictional
component is shown as a dashpot consisting of a plunger moving through
a viscous fluid.

If the mass is displaced, it will experience three forces: a spring
force, an inertial force, and a damping force. The differential equation
of motion is given by equation 18.

$$ma + Bv + kx = 0 \qquad\qquad 18$$

Figure 15

Damped Free System

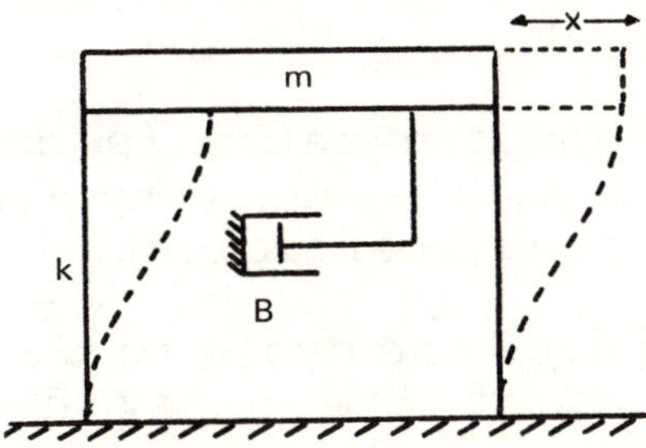

Recognizing that acceleration is the second derivative of position and
that velocity is the first derivative of position enables us to write
equation 19.

$$mx'' + Bx' + kx = 0 \qquad\qquad 19$$

Or,
$$x'' + (B/m)x' + (k/m)x = 0 \qquad\qquad 20$$

Since the natural frequency of the undamped case can be written in terms
of k and m, we may write equation 20 as equation 22.

$$\omega = \sqrt{(k/m)} \qquad\qquad 21$$

$$x'' + 2\xi\omega x' + \omega^2 x = 0 \qquad\qquad 22$$

For moderate damping, the solution of this differential equation is the
decaying sinusoid shown in figure 16. This sinusoid fits within the
exponential decay envelope specified by equation 23.

$$x = e^{-\xi\omega t} \qquad\qquad 23$$

Figure 16

Damped Free Oscillation

(moderate damping)

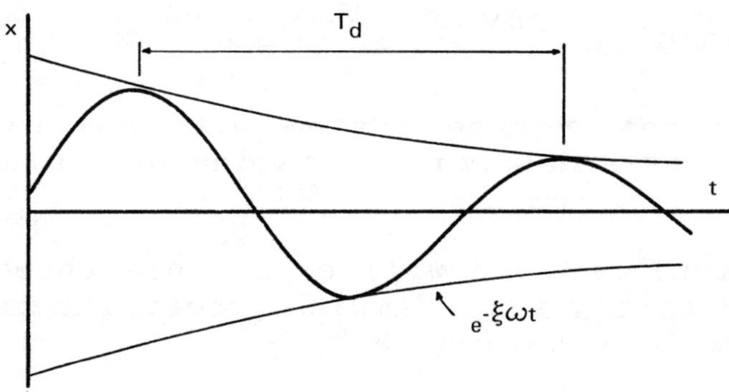

ξ is an important parameter known as the damping ratio. This damping
ratio is given by equation 24.

$$\xi = \frac{B}{2m\omega} \qquad\qquad 24$$

It is apparent that the damping ratio is calculated by dividing the
actual damping coefficient by the quantity $(2m\omega)$. This denominator is
known as the critical damping coefficient.

The ratio of the amplitude from one cycle to the next is known as the decay
decrement.   The logarithm of the decay decrement is known as the
logarithmic decrement.

$$\log_e(x_n/x_{n+1}) = \frac{2\pi(\xi)}{\sqrt{1-(\xi)^2}} \qquad\qquad 25$$

The value of B (and hence of ξ) will determine the performance of the
system.   There are three cases.

case 1: If the damping ratio is less than one, the oscillations are
underdamped.   The motion will be oscillatory with a diminishing
amplitude as shown in figure 16. The frequency of oscillation will
not be the same as for the undamped case since the damping slows down
the movement of the mass. The frequency and period of the damped
system are given by equations 26 and 27.

$$\omega_d = \omega\sqrt{1 - (\xi)^2} \qquad\qquad 26$$

$$T_d = 2\pi/\omega_d \qquad\qquad 27$$

case 2: If the damping ratio is greater than one, the motion will die out at a rate determined by the damping ratio. This overdamped case is illustrated in figure 17.

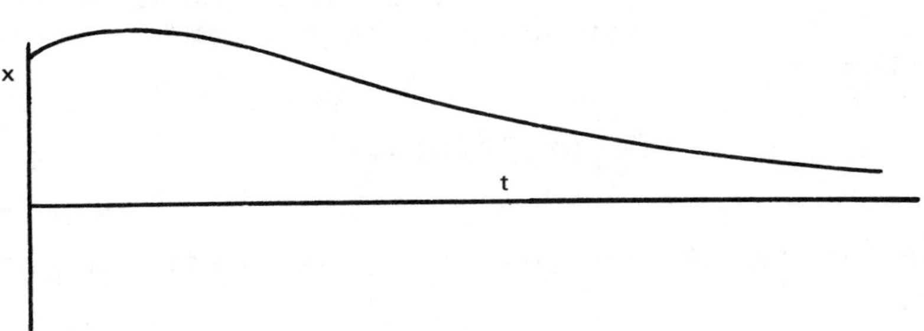

Figure 17

Overdamped Motion

case 3: If the damping ratio is equal to one, the system is said to be critically damped. There is no overshoot and the vibrations die out the fastest of the three cases. The damping coefficient required to achieve critical damping is known as the critical damping coefficient.

$$B_{critical} = 2m\omega \qquad\qquad 28$$

Figure 18

Critically Damped Motion

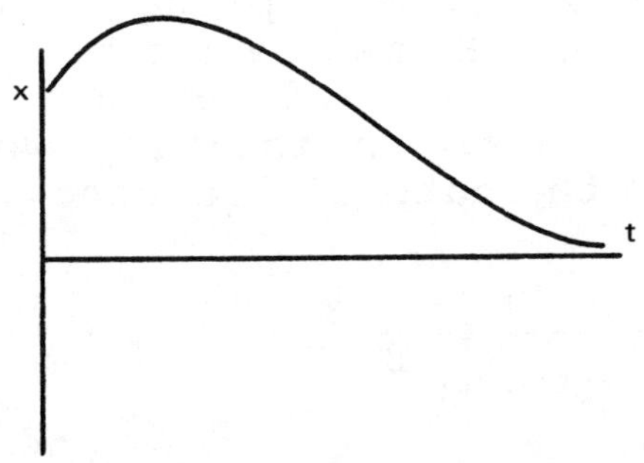

C. Undamped Forced Systems

A forced system is one which is acted upon by a force which supplies energy on a regular or irregular basis. An example of a regularly forced system is a floor which supports a motor. The motor imparts to the floor a loading with a frequency proportional to its rotational speed. An example of an irregularly forced system is a structure acted upon by wind or seismic forces. In this latter case, there is no regularity to the forces applied to the structure.

If the forcing function is written as F(t) to indicate its dependence on time, the differential equation of motion is

$$mx'' + kx = F(t) \qquad\qquad 29$$

The response of the system to the applied force depends on the nature of the forcing function. For the time being, assume that the forcing function has a magnitude of P and a periodic frequency of ω_f. (Notice that the frequency of the forcing function does not have to be the same as the natural frequency of the system.) Then, we may write the forcing function as

$$F(t) = P\sin(\omega_f t) \qquad\qquad 30$$

The solution of the resulting differential equation has the following form:

$$x(t) = A_1\cos\omega t + A_2\sin\omega t \qquad\qquad 31$$

Use of trigonometric substitutions reduces equation 31 to 32.

$$x(t) = A\sin\omega_f t \qquad\qquad 32$$

The constant A is given by equation 33.

$$A = (MF)\left(\frac{P}{k}\right) \qquad\qquad 33$$

The quantity (P/k) in equation 33 is just the deflection that would occur if a constant force P acted on the spring. The quantity (MF) is known as the magnification factor. MF may be greater or less than one depending on the forcing frequency. The magnification factor is graphed in figure 19. Notice the MF increases to infinity when the forcing function frequency coincides with the natural frequency. This situation is known as resonance.

$$(MF) = \frac{1}{1 - (\omega_f/\omega)^2} \qquad\qquad 34$$

If the forcing function is a short impulse, an exact solution may be obtained from the differential equation of motion by using Laplace transforms. An impulse load may result from a projectile impact, a sudden wind gust, or a short-duration seismic tremor. Assuming that the length of the impulse is much less than the natural period of the structure, the approximate response is given by equation 35.

$$x(t) \approx \frac{\int F\,dt}{m\omega}\sin\omega t \qquad\qquad 35$$

The same response can be expected from all short duration impulses which have the same values of $\int F\,dt$.

If the structure is acted upon by an arbitrary loading (not necessarily with a duration less than the natural period of vibration) the loading

Figure 19

Undamped Magnification Factor

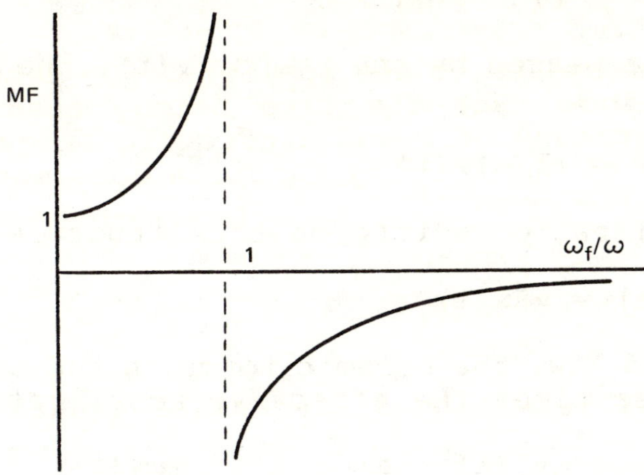

may be treated as a series of short impulses.   The response to a series
of short impulses is given by equation 36.

$$x(t) = \int_0^t \frac{F(\tau)}{m\omega}\, \sin\omega(t-\tau)d\tau \qquad\qquad 36$$

Equation 36 is known as Duhamel's integral.   Duhamel's integral is
essentially the application of superposition to a series of pulses, each
ending at time τ.

D. Damped Forced Systems

The differential equation of motion for a damped forced system is

$$mx'' + Bx' + kx = F(t) \qquad\qquad 37$$

The solution to this equation depends on the form of F(t).   Solutions to
the sinusoidal forcing case are readily available in systems and dynamics
textbooks.   If the forcing function is an impulse, the response is

$$x(t) = \frac{\int F\,dt}{m\omega_d}\, e^{-\xi\omega t}\sin\omega_d t \qquad\qquad 38$$

If the structure is subjected to an arbitrary loading, such as would
occur in an earthquake, the response can be determined by writing
Duhamel's integral for the damped case.

$$x(t) = \int_0^t \frac{F(\tau)}{m\omega_d}\, e^{-\xi\omega(t-\tau)}\sin\omega_d(t-\tau)dt \qquad\qquad 39$$

10. Earthquake Response of SDF Systems

Duhamel's integral for damped motion applies to a structure which
experiences a displacement due to the application of a force.   During an
earthquake, it is the ground that moves, not the structure.   Therefore,
Duhamel's integral can be rewritten in terms of the ground acceleration,
x''_g.   That is,

$$x''_g(\tau) = F(\tau)/m \qquad\qquad 40$$

If it is also assumed that ω and ω_d are approximately equal, then Duhamel's integral becomes

$$x(t) = \frac{1}{\omega}\int_0^t x_g''(\tau)e^{-\xi\omega(t-\tau)}\sin\omega(t-\tau)d\tau \qquad 41$$

If the integral is replaced by the symbol $v(t)$, the earthquake response is given by equation 42.

$$x(t) = (1/\omega)v(t) \qquad 42$$

The effective acceleration acting on the structure is

$$x_e''(t) = \omega^2 x(t) \qquad 43$$

From Newton's second law, the seismic force on the structure is found by multiplying the mass times the effective acceleration. Therefore,

$$V(t) = mx_e''(t) = m\omega^2 x(t) = m\omega v(t) \qquad 44$$

The last form of $V(t)$ is the most important since it indicates that the force induced by an earthquake (the 'base shear') can be found from the structure's mass and natural frequency of vibration.

The application of equation 44 depends on being able to evaluate the integral $v(t)$. Many numerical integration methods are available to do so. However, when the ground motion is not known in advance, such an integration is not possible. Since earthquake motions are not known in advance, it is usually acceptable to work with an estimate of the maximum value of $v(t)$.

The maximum value of $v(t)$ is known as the spectral velocity, S_v. (This is also known as the spectral pseudo-velocity because it does not correspond exactly to the maximum velocity due to the approximations made in the analysis.)

The maximum earthquake displacement is the spectral displacement.

$$x_{max} = S_d = S_v/\omega \qquad 45$$

The spectral acceleration is

$$S_a = \omega S_v = \omega^2 S_d \qquad 46$$

Therefore, the maximum effective earthquake force (base shear) is

$$V_{max} = mS_a \qquad 47$$

S_v can be evaluated for different natural periods of oscillation. This will yield a plot of S_v versus ω (or T). Such a plot is known as a response spectrum. It should be remembered that such a response spectrum depends on the nature of the forcing function and the damping ratio.

Figures 20, 21, and 22 show spectral accelerations, velocities, and displacements for specific earthquakes with 5% structural damping. These

values are not presented as graphical solution aids (since they apply to
a specific earthquake at a specific location) but rather to show the
general shape of response spectra.

Figure 20

Spectral Accelerations

El Centro Earthquake, May 1940, North-South Directions, 5% Damping

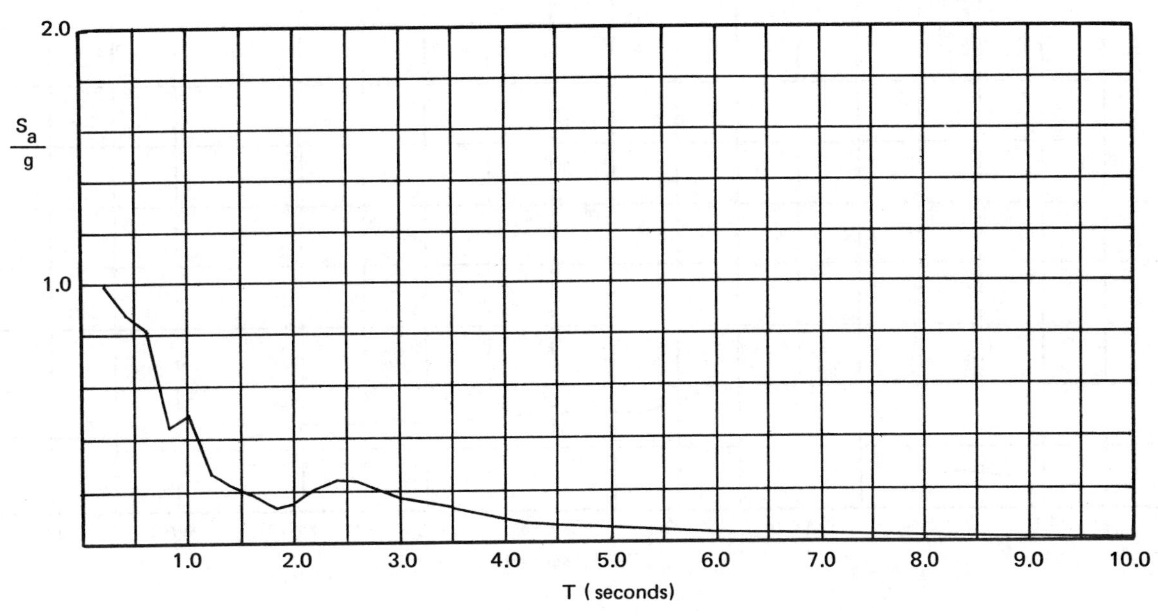

Figure 21

Spectral Velocity

El Centro Earthquake, May 1940, North-South Directions, 5% Damping

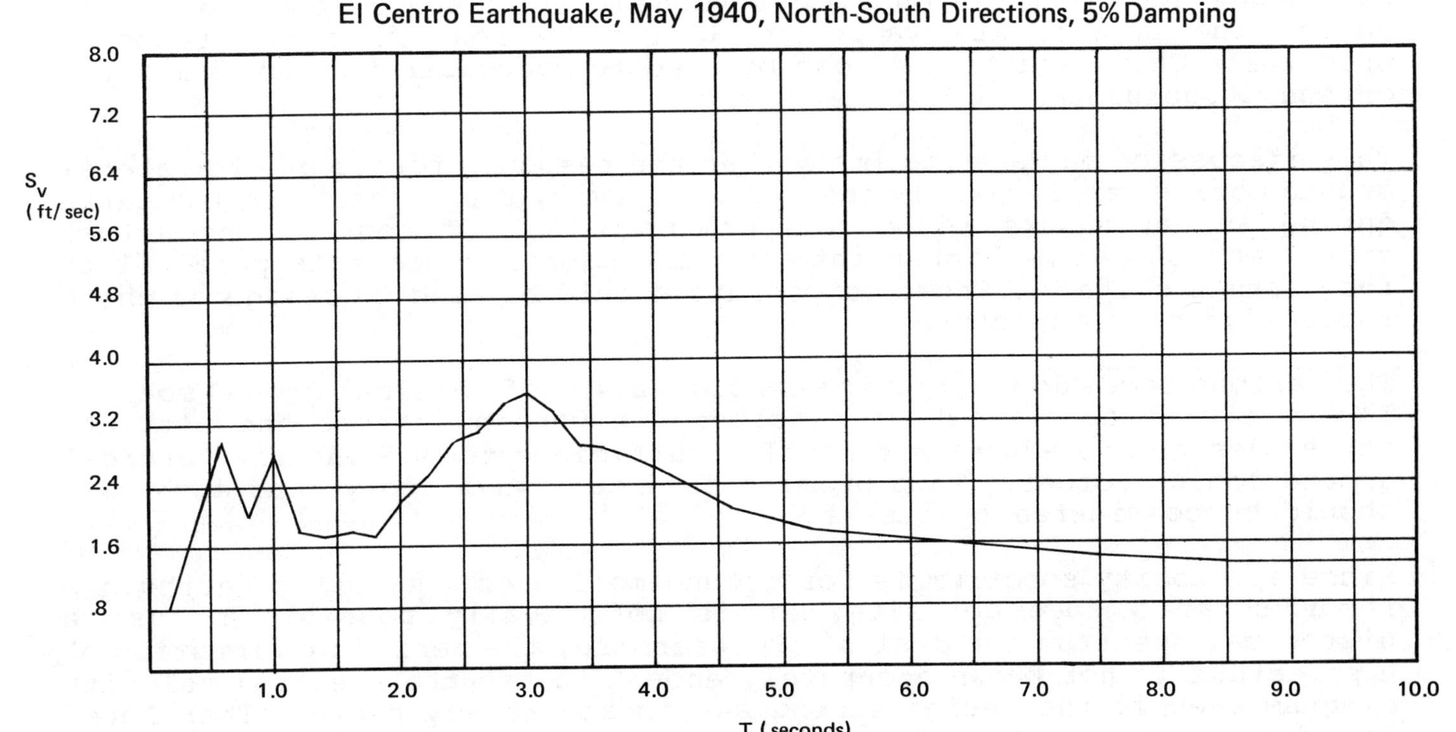

Figure 22

Spectral Displacement

El Centro Earthquake, May 1940, North-South Directions, 5% Damping

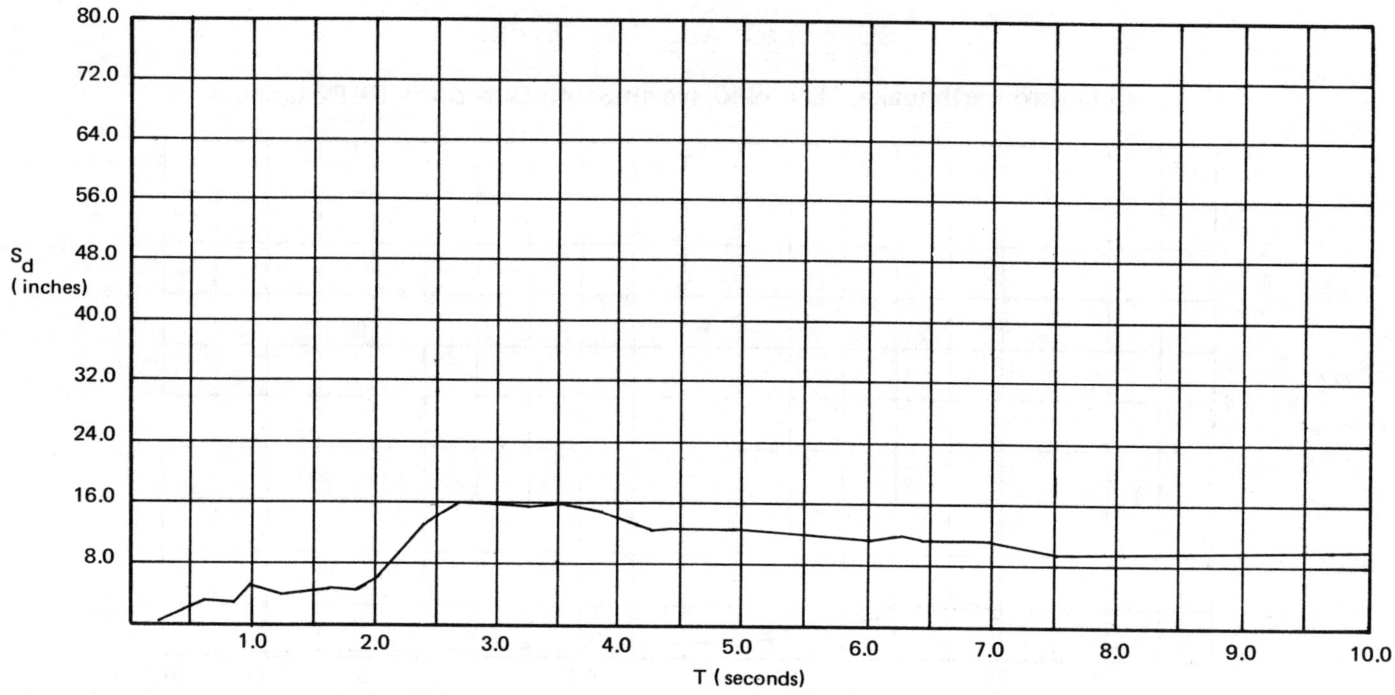

The method of evaluating the seismic forces on a structure should now be apparent. Once the structure is identified, its mass, damping ratio, and natural period of vibration are determined. For a specific motion of the earth, the appropriate value of S_a can be read from the response spectrum. Then, equation 47 can be used to calculate the maximum force on the structure.

This discussion may seem to imply that the response forces due to lateral ground motions apply to only one axis. However, the movement of the earth during an earthquake is not one-dimensional. Of course, the actual ground motion can be broken into two orthogonal components parallel to the earth's surface. These components should be used to design or check the design of a structure.

The various codes do not require consideration of vertical ground motion. The safety factors which are included in working stress and ultimate stress design procedures are usually sufficient to provide for vertical ground motion forces. This premise may not always apply, and each case should be considered by itself.

Since a velocity spectrum is for ground motion of a specific earthquake recorded at a specific site, it is not usually possible to use a historical spectrum for design. Furthermore, the period of vibration of a structure is not known accurately enough to use the spectral velocity diagram even if the design earthquake is accurately known. Therefore, an average spectrum based on an envelope of actual cases as shown in

figure 23 will generally be more appropriate. Such an average spectrum
is known as a design spectrum.

Figure 23

Average Response Design Spectra for the

1940 El Centro Earthquake

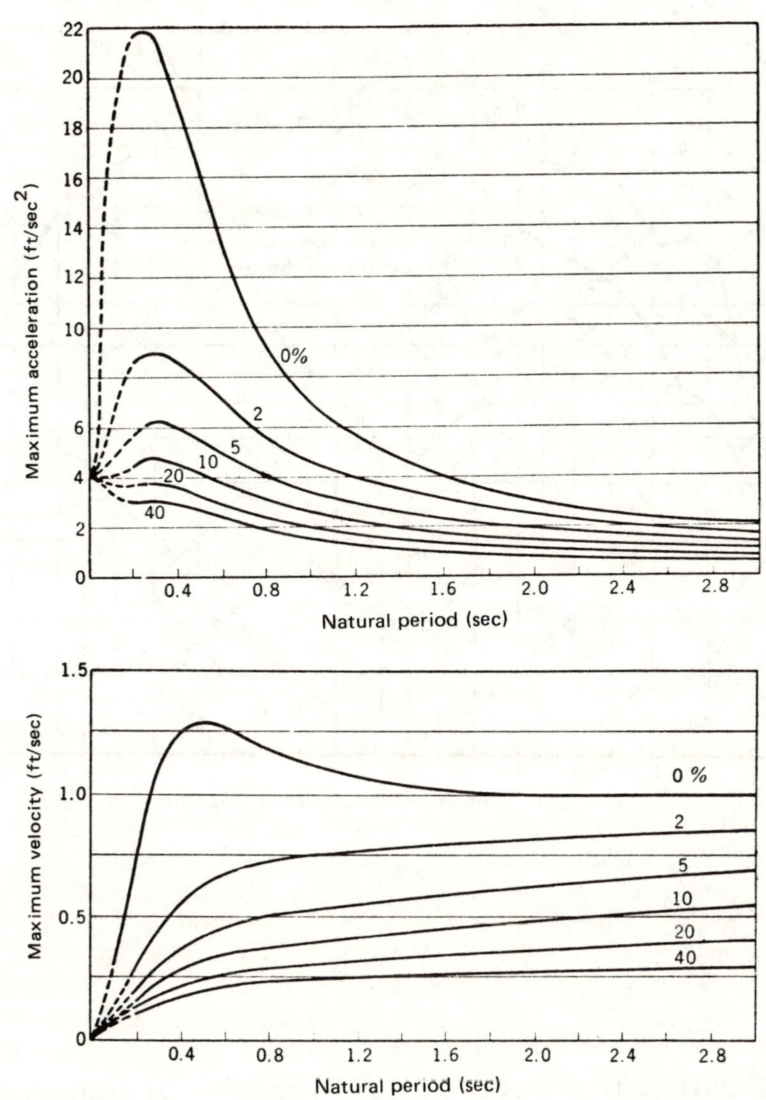

Because acceleration, velocity, and displacement are all related, all
three may be shown by a single curve on a graph which has three different
scales. Such a graph is usually done on a logarithmic scale and is known
as a log tripartite plot. Although log tripartite plots provide a
convenient method of graphing complex data, they may have the effect of
concealing important features.

Figure 24

Typical Log Tripartite Plot

El Centro Earthquake (elastic)

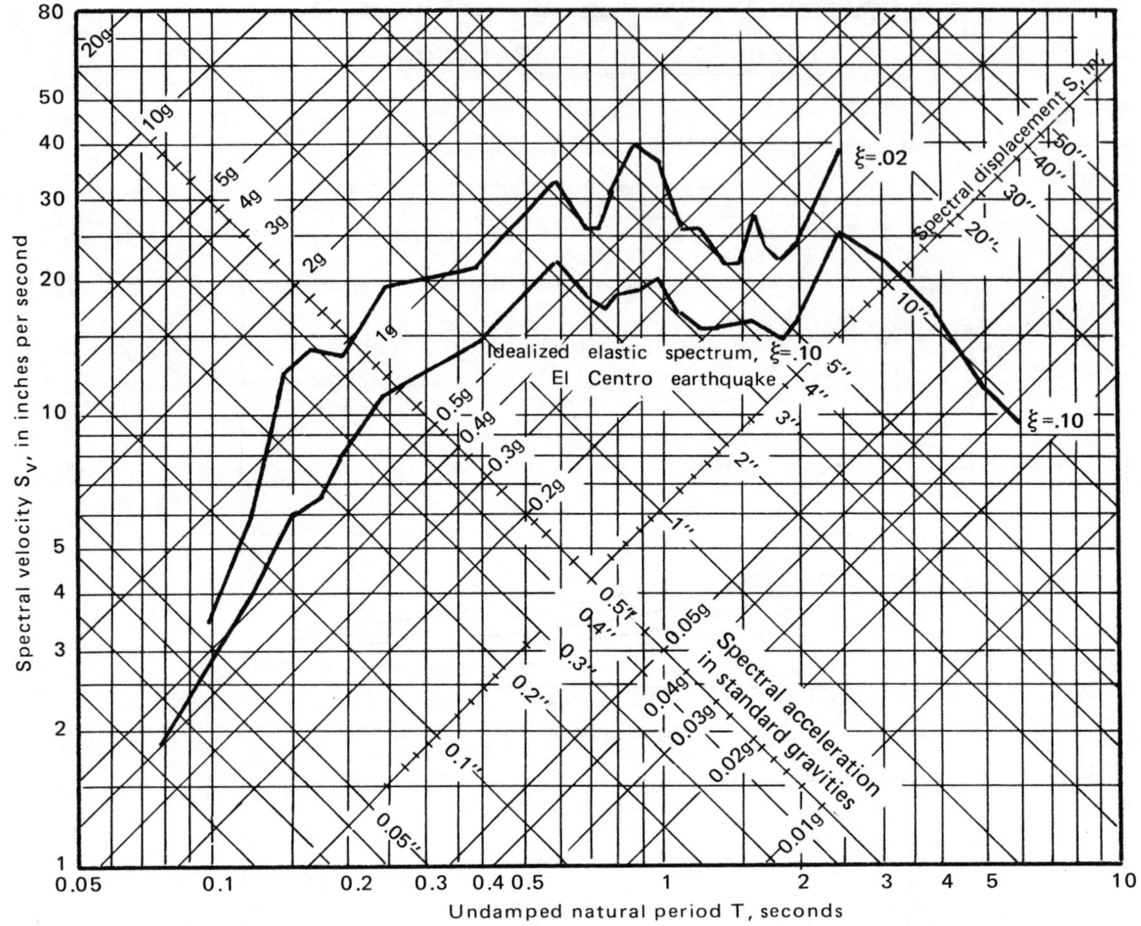

Evaluation of all available data on California earthquakes has resulted in the following observations:

(1)   None of the spectra exhibited resonance. Although a true resonant response (an infinite magnification factor) is theoretically possible, it can be considered highly unlikely.

(2)   Typical ground motions for California earthquakes show S_a peaks in the .2 to .5 second period range.

(3)   The peak S_a for a 10% damped system is approximately 2 to 2.5 times the maximum ground acceleration.

Example 3

The primary support of a drill press weighing 100,000 pounds is the structural steel bent shown below. Assume 5% damping and calculate the base shear and spectral responses for the 1940 El Centro N-S earthquake. Neglect the column weights.

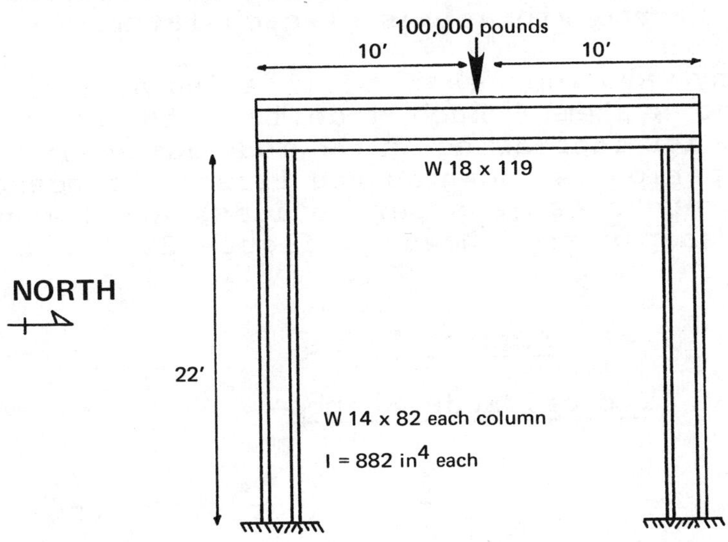

Solution

The mass of the drill press and beam is

$$[100,000 + (20)(119)]/32.2 = 3180 \text{ slugs}$$

Considering the columns to be fixed at both ends, the deflection per pound of lateral loading is

$$\frac{L^3}{12EI} = \frac{[(22)(12)]^3}{(12)(3 \text{ EE}7)(2)(882)} = 2.9 \text{ EE-5 in/lb}$$

$$= 2.42 \text{ EE-6 ft/lb}$$

Then, $k = 1/(2.42 \text{ EE-6}) = 4.13 \text{ EE5}$

The period is $T = 2\pi\sqrt{3180/(4.13 \text{ EE5})} = .55 \text{ seconds}$

From figures 20, 21, and 22, the approximate spectral responses with 5% damping are

$$S_d = 2 \text{ in}$$

$$S_v = 2.40 \text{ ft/sec}$$

$$S_a = .85g = (.85)(32.2) = 27.4 \text{ ft/sec}^2$$

The base shear is $V = ma = (3180)(27.4) = 87130 \text{ lb}$

11. Multiple Degree of Freedom Models

A. Analytical Methods

A structure with several layers of mass (such as a building with several floors) does not behave like a SDF system. Such a structure must be evaluated as a multiple degree of freedom (MDF) system whose vibration will be a combination of the vibrations of each layer.

A MDF system has as many ways (modes) of oscillating as there are layers (lumped masses) in the system. Each mode has its own frequency of vibration, the frequencies increasing as the mode number increases. The mode with the longest period is known as the first or fundamental mode. Modes with greater frequencies (shorter periods) are known as higher modes. Typical mode shapes are shown in figure 25 for a five-story building.

Figure 25

Typical Mode Shapes

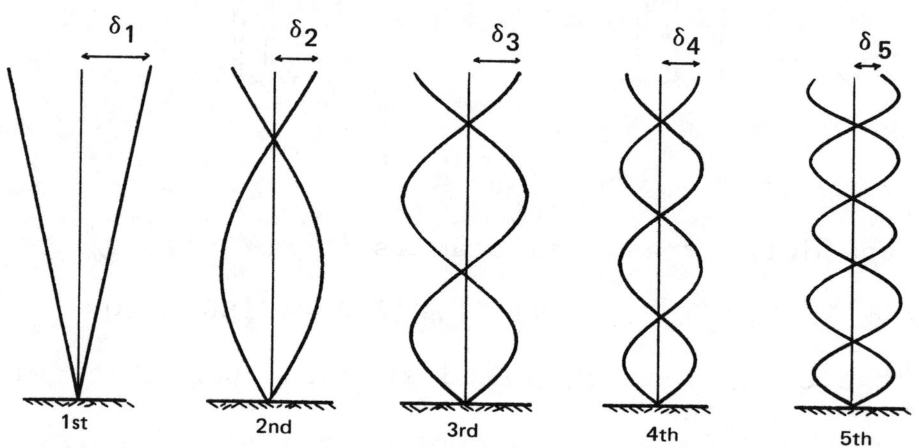

A knowledge of the mode of vibration is necessary if the structure's response is to be determined. Computer matrix analysis can be used to evaluate multi-story buildings based on the independent SDF performance of each floor. As was done with SDF systems, considerable simplification can be achieved by working only with the maximum deflections. However, even this simplification requires a probabilistic analysis since the modal maxima do not necessarily occur at the same time nor do they have the same sign when they do occur.

Various approximation formulas are used, but the most common is the sum-of-squares approximation. If the maximum displacements (δ_i) are known for the first n modes, equation 48 usually gives a conservative estimate of the top story displacement. (Actually, this method can be used for any story whose mode shapes are known.)

$$\Delta = \sqrt{\Sigma \delta_i^2}$$

48

Since most of the energy of vibration is absorbed by the first three to six mode shapes, no other modes need to be evaluated. This results in considerable savings in computational effort.

The periods of the various modes decrease rapidly from the fundamental (first) mode. For typical buildings with the same general plan view for all stories, the decrease is in the order of 1, (1/3), (1/5), (1/7), (1/9), etc. The participation decreases even more quickly than that, which is the reason that only the first few mode shapes need to be considered. (Participation is defined as the percent of the total building mass which acts in any particular mode.) It is apparent that the fundamental mode dominates.

Since the fundamental mode dominates, the response spectra for acceleration, velocity, and displacement are similar to SDF systems, particularly in the intermediate period ranges. For short periods, the MDF response is less than that for the SDF system. For periods exceeding 1 second, the MDF response may slightly exceed the SDF response.

Example 4

Compute the natural frequencies of the MDF system shown below.

Solution

If x_1, x_2, and x_3 are the displacements of masses 1, 2, and 3 (measured with respect to the equilibrium position), the spring forces on each lumped mass are:

The above freebodies are not in equilibrium. (This is particularly obvious for mass 3.) Using D'Alembert's Principle of Dynamic Equilibrium, (ma) terms are added to account for the inertial forces. From equation 46,

$$ma = m\omega^2 x$$

Then, the equilibrium conditions for the masses are:

mass 1: $\quad [m_1\omega^2 - (k_1+k_2)]x_1 + k_2x_2 = 0$

mass 2: $\quad k_2x_1 + [m_2\omega^2 - (k_2+k_3)]x_2 + k_3x_3 = 0$

mass 3: $\quad k_3x_2 + [m_3\omega^2 - k_3]x_3 = 0$

Since all m's and k's are known, this can be written in the matrix form below.

$$\begin{bmatrix} (\omega^2-200) & (100) & (0) \\ (100) & (\omega^2-200) & (100) \\ (0) & (100) & (.5\omega^2-100) \end{bmatrix} \begin{bmatrix} x_1 \\ x_2 \\ x_3 \end{bmatrix} = \begin{bmatrix} 0 \\ 0 \\ 0 \end{bmatrix}$$

Disregarding the trivial solution, the large matrix must have a determinant of zero. Setting the determinant equal to zero gives the following equation:

$$\omega^6 - (600)\omega^4 + (90,000)\omega^2 - 2,000,000 = 0$$

This has solutions of: $\omega_1 = 5.18$, $\omega_2 = 14.14$, and $\omega_3 = 19.32$.

These are the natural angular frequencies (radians/second) for this system.

◆

Once the natural frequencies are known, the mode shapes can be determined. The mode shape is a plot of the maximum excursion (from the equilibrium position) of each lumped mass. The mode shapes are usually normalized so that the following relationship holds.

$$\sum_i m_i\phi_i^2 = 1 \qquad\qquad 49$$

ϕ is known as the mode shape factor. The mode shapes are easily found by substituting ω into the equilibrium conditions. This process may be facilitated by initially assuming a value for one of the maximum excursions.

Example 5

Find the first mode shape of the system evaluated in example 4.

Solution

For the first mode, $\omega^2 = (5.18)^2 = 26.8$. So, the equilibrium requirement for mass 1 is

$$[(26.8 - 200)]x_1 + 100(x_2) = 0$$

Assuming $x_1 = 1$, then, $x_2 = 1.732$

Similarly, the equilibrium condition for mass 3 is

$$(100)(1.732) + [(.5)(26.8) - 100]x_3 = 0$$

Or, $x_3 = 2$

The mode shape is shown below. It can be normalized by dividing the maximum excursions by $\sqrt{6}$.

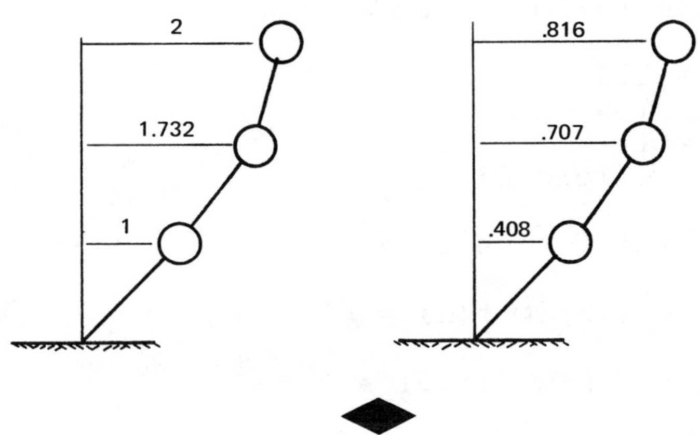

B. Approximate Methods

There are computer algorithms that can perform a dynamic analysis of a MDF system. However, use of the computer may not be economical for systems with less than 10 degrees of freedom. However, the solution of such systems by hand is also out of the question. Luckily, several approximate procedures are available for determining the natural frequencies and mode shapes of MDF systems.

Most of these approximate procedures are based on the "Rayleigh method." Although there are many variations of the Rayleigh method, the basic approach is to set the maximum kinetic energy equal to the maximum potential energy. Since the mode shape is not known, it has to be assumed to use the Rayleigh method. Even a very crude assumption will give a reasonable value for the frequency of the fundamental mode.

The Stodola and Holzer methods are iterative processes used to find the principal modes and natural frequencies of free undamped SDF systems. The Stodola method uses the following steps:

step 1:   Assume a deflection for each lumped mass.

step 2:   Compute the inertial forces using equation 46. That is, the inertial force is $m\omega^2 x$.

step 3:   Compute the spring forces as the sum of the inertial forces acting on the springs.

step 4:   Compute the spring deflections.

step 5:   Calculate the mode deflections from the deflections. If they match the assumed deflections from step 1, the problem is solved. If they do not match, the spring deflections from this step can be used to start the procedure again.

Example 6

Use the Stodola method to solve example 4.

Solution

step 1:   Assume the following mode shape.

$$x_3 = 2.0$$
$$x_2 = 1.5$$
$$x_1 = 1$$
$$x_0 = 0 \text{ (ground)}$$

step 2:   The inertial forces are:

$$F_{I3} = (.5)(\omega^2)(2) = \omega^2$$

$$F_{I2} = (1)(\omega^2)(1.5) = 1.5\omega^2$$

$$F_{I1} = (1)(\omega^2)(1) = \omega^2$$

step 3:   The spring forces:

$$F_{s3} = F_{I3} = \omega^2$$

$$F_{s2} = F_{I2} + F_{I3} = 2.5\omega^2$$

$$F_{s1} = F_{I1} + F_{I2} + F_{I3} = 3.5\omega^2$$

step 4:   The spring deflections are:

$$x_{s3} = F_{s3}/k_3 = .01\omega^2$$
$$x_{s2} = F_{s2}/k_2 = .025\omega^2$$
$$x_{s1} = F_{s1}/k_1 = .035\omega^2$$

step 5:   Dividing by ω^2, the new relative mode deflections are:

$$x_1 = x_{s1} = .035$$

$$x_2 = x_{s1} = x_{s2} = .060$$

$$x_3 = x_{s1} + x_{s2} + x_{s3} = .075$$

Dividing by .035, the mode shape is:

$$x_3 = 2.14$$
$$x_2 = 1.71$$
$$x_1 = 1$$

These values can be used to start the procedure all over again. Eventually, the values from steps 1 and 5 will agree.

At the end of this iteration, the best estimate of the natural frequency can be found from

$$(1 + 1.71 + 2.14) = (.035 + .060 + .075)\omega^2$$

Or, $\omega = 5.34$ radians/sec

◆

12. Seismic Codes

A 'code' is a set of rules that is adopted by a body empowered to enforce the code. The mere publication of a set of guidelines does not constitute a code. Although the Structural Engineers Association of California (SEAOC) publication entitled "Recommended Lateral Force Requirements and Commentary" falls into this second category, the recommendations are commonly referred to as the SEAOC code.

The SEAOC code is the most detailed and scientific seismic code. The SEAOC code is used almost without modification in the Uniform Building Code (UBC) as Chapter 23. Therefore, the UBC is essentially equivalent to the SEAOC code. Relevant portions of the SEAOC code are also used in the American Institute of Timber Construction's Timber Construction Manual and the publications of the Portland Cement Association and the Masonry Institute.

There are no special seismic provisions in the American Institute of Steel Construction Manual of Steel Construction. Important seismic provisions of the American Concrete Institute's "Building Code Requirements for Reinforced Concrete" (ACI 318-77) are discussed later in this document.

13. Seismic Design Steps

There are basically six different steps that must be taken in a seismic design or analysis problem. These steps are presented below in general terms. Many specific provisions (such as building periods) are covered in the SEAOC code.

Step 1: Determine the Design Earthquake

Two different types of ground motions may be defined. The 'probable maximum earthquake' is the largest earthquake that has a significant probability of occurring within the lifetime of the building. Since the risks are very great, the probability does not have to be very large to be significant.

The 'maximum credible earthquake' is the maximum possible earthquake based on available knowledge about the building site. The maximum possible earthquake is difficult to evaluate since no one knows what maximum possibilities exist. However, energy requirements seem to limit the maximum worldwide Richter magnitude to 9.0. In California, the maximum is generally considered to be 8.5.

A prediction of the design earthquake must make use of available statistical data on past seismic events. Predictions should also consider the performance of neighboring structures and the design earthquakes used in their structural development.

Step 2: Determine the Damping Ratio

The SEAOC code does not specify damping ratios due to lack of experimental data on that subject. The selection is further complicated by the variation (increase) in damping ratio during swings of large amplitude. However, available data on actual structures seems to suggest the following values for the damping ratio: steel braced frames, 2%; steel moment resisting frames, 5%; concrete shear walls, 5% - 10%; wood frames, 15%. There is little evidence to expect the damping ratio to exceed 15%.

Step 3: Determine the Structure's Ductility

Ductility is the capability of a structure to distort during yielding without failure or loss of strength. During a seismic event, a ductile structure will be able to absorb oscillatory energy even after the structure has experienced local or total yielding.

The SEAOC code does not specify acceptable stress levels. However, the expected magnitude of the seismic loads makes it necessary to accept some inelastic yielding during an earthquake. The most common method of accounting for ductility is to factor down the elastic response spectra by a chosen ductility factor.

Ductility of a joint or a member may be specified by the ductility factor. Unfortunately, there are a number of definitions of this ductility factor.

definition 1: the ratio of displacement (strain) required to absorb the energy of an earthquake to the displacement (strain) at yielding.

definition 2: the ratio of the maximum expected strain to the strain at the yield point.

definition 3: the ratio of the angular rotation of a member or joint at ultimate failure to the angular rotation at the onset of yielding

definition 4: the ratio of the story shear caused by seismic acceleration to the maximum elastic story shear

These definitions are not interchangeable, although they are related. A ductility factor (based on the third definition) up to 8.00 may be tolerated in certain types of structures without collapse. The minimum ductility possessed by modern structures seems to be about 2.5. Desireable levels vary, although 6 to 8 is usually sought for steel frames, and 4 to 6 for concrete frames.

Step 4: Determine the Building Period

It was previously shown that the natural period of a simple harmonic oscillator is given by equation 50.

$$T = 2\pi\sqrt{(m/k)}$$

From Hooke's law, $k = F/\Delta$. Also, the mass is (W/g). Therefore, the period can be written as

$$T = 2\pi\sqrt{\frac{W\Delta}{gF}} \qquad\qquad 51$$

If the weight is expressed as a fraction $(1/C_F)$ of the applied force, this can be written as equation 52 in which Δ is expressed in inches. C_F is known as the lateral force coefficient. (C_F) is equal to one if the entire building is used as the horizontal dynamic force. This could only be approximated with short and stiff buildings.

$$T = .32\sqrt{(\Delta/C_F)} \qquad\qquad 52$$

$$C_F = \frac{V}{W} = ZICS \qquad\qquad 53$$

Even though the above analysis was for a SDF system, the period formula can be extrapolated to MDF systems. For any mode in a MDF system, the mode period constant is C_T. The period for that mode is

$$T = C_T\sqrt{\Delta/C_F} \qquad\qquad 54$$

A typical value of C_T is .25 for MDF, constant drift, moment-resisting frame systems oscillating in the fundamental mode. The use of equation 54 assumes the displacement is calculated from the code drift limitations. If the displacement is the actual maximum, C_F is taken as 1.0.

The value of Δ may have to be estimated from the story drift ratios. A story drift ratio is the ratio of story displacement to story height (floor to floor distance). The story displacement is measured at the story ceiling relative to the story floor. It is not measured relative to the ground. Therefore, Δ would be equal to the sum of the story displacements.

The Uniform Building Code and SEAOC code both specify that T may be calculated as $(.10)N$ where N is the number of stories. For a constant drift building, N and Δ are proportional. Since the theoretical period equation requires the period to be related to the square root of the roof displacement, $(.10)T$ is a linear approximation to a non-linear curve. The approximation is good for buildings in the 40-story range, but it is very poor for short buildings.

The alternate period equation (SEAOC code 1-3A) is based on recorded periods on shear wall buildings.

Of course, there is nothing stopping a structural designer from using exact analytical methods of evaluating T. In fact, such an analysis should be performed on special structures such as towers and chimneys.

<u>Step 5: Determine the Dynamic Response to the Ground Motion</u>

There are three methods of determining the dynamic response. All of these are consistent with most building codes and city building department requirements.

<u>method 1</u>: The SDF spectra may be used to determine the response for the principal modes of vibration. Then, the SDF

modal responses may be combined on the basis of probability and participation. Response may also be arbitrarily reduced to account for inelastic response.

method 2: A computer may be used to simulate ground motion and to directly determine the MDF response of the structure by combining modal responses for every interval of time at each level.

method 3: A computer may be used to numerically integrate the design earthquake to determine the dynamic response.

Step 6: Distribute the Generated Forces

In the SDF system, the inertial seismic load is applied to the single supported mass. For MDF models, the seismic load is distributed non-uniformly along the height of the building.

The total horizontal seismic shear at the base of a structure is known as the base shear. Base shear is the sum of the inertial forces for each mass in the structure. The inertial forces (calculated as the mass times the acceleration) are added algebraically.

Since the inertial accelerations experienced by a building varies linearly with height above the earth's surface, the story shears will vary linearly with height. (This assumes all story masses are the same.) Thus, the distribution of the seismic load is triangular. In addition, the SEAOC code specifies that a portion of V (known as F_t) be applied at the roof.

Figure 26

MDF Loading Diagram

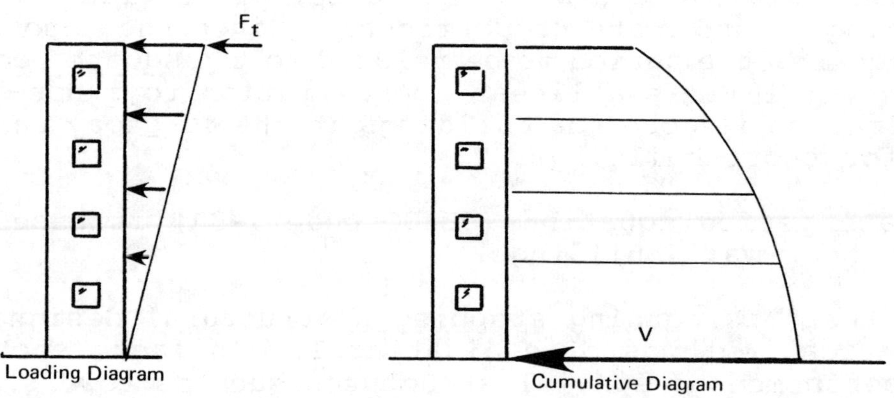

The base shear is specified in the code as equation 55.

$$V = ZIKSCW = C_F KW$$

55

Table 6

Values of Z

seismic zone	Z
1	3/16
2	3/8
3	3/4
4	1

I is an occupancy importance factor. It is given in table 7.

Table 7

Values of I

type of facility	I
Essential facilities such as hospitals, fire and police stations, and disaster control centers	1.5
Any building whose primary purpose is to hold more than 300 people in one room	1.25
All other facilities	1.00

The value of K requires knowledge of the type of construction. The lowest value of K(.67) is assigned to buildings with ductile moment-resisting space frames. This type of frame remains stable independently of failures of other structural elements such as walls.

A ductile moment-resisting frame possesses joints that achieve ductile failure without buckling or major distortion. In steel construction, for example, extensive use is made of type 1 connections. These connections transmit column moments directly to the beams and girders causing the beams and girders to carry some of the applied moments. Without such moment-resisting joints, it is likely that the columns would fail (for example, by web crippling) without stressing the beams and columns to any great extent. Therefore, the beam-column connection must be able to transmit without failure a moment equal to the plastic capacity of the beam.

A building constructed as a box (K = 1.33) relies on walls to carry vertical and lateral loads. Since walls are generally constructed of wood framing or concrete, their failure is non-ductile.

Braced frames (K = .80 and 1.0) possess joints incapable of transmitting any significant moments. Thus, the joints of a braced frame can be considered to be pin-joints. In steel design, simple frame connections are known as type 2 connections.

Figure 27

Type 1 Joints

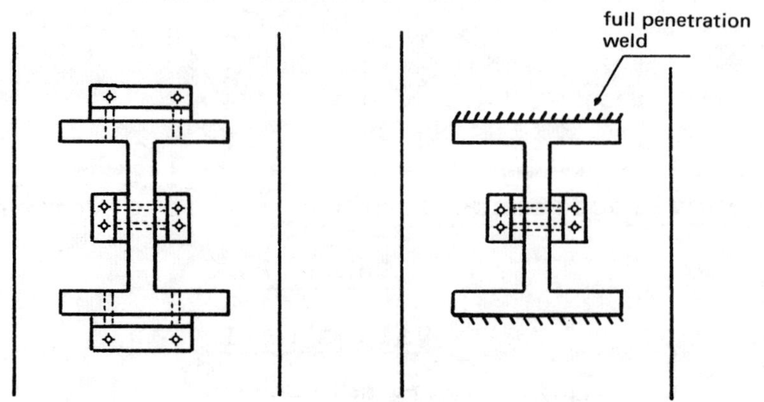

Figure 28

A Type 2 Joint

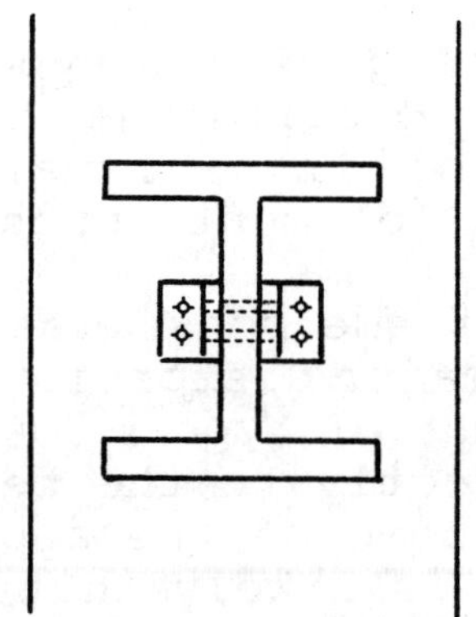

The additional SEAOC code requirements listed below are also relevant to the choice of K.

(a) Values of .67 and .8 for K may only be used with buildings having ductile moment-resisting space frames.

(b) Values of 1.00 and 1.33 are limited to buildings of less than 160 feet in height. K must be either .67 or .80 for buildings higher than 160 feet since buildings of this height must possess a ductile moment-resisting space frame.

(c)   If K = 1.33 is required in one direction (such as North-South) it will also be required in the other direction (such as East-West).

(d)   For buildings less than 160 feet high, any combination of K = 1.00, .80, and .67 in the two directions may be used.

Formulas for C and S (which depend on T) are given in the SEAOC code. Since C is related to the square root of T, extreme accuracy in T is not needed. In some cases, trial and error solutions for the related quantities of C, S, and T are needed. This is simplified by the knowledge that S has a narrow range, generally between 1.0 (for rock or soil) and 1.5. (S is limited by the SEAOC code to values not less than 1.0)

Example 7

A 6000 kip, 7-story office building is being planned for downtown San Francisco. Estimate its first-mode natural period of oscillation if an 85-foot tall ductile moment-resisting frame is to be used. The site period is estimated to be 1 second.

Solution

The SEAOC code only has 3 formulas for calculating the period of vibration. The first (equation 1-3) cannot be used because the base shear is unknown and therefore cannot be distributed. Equation 1-3A is for shear wall construction. Equation 1-3B is for buildings with approximately 40 stories.

Therefore, equation 54 must be used by assuming maximum drift.

$$T = .25\sqrt{\Delta/C_F}$$

The SEAOC code limits the story drift to .5%. So, the maximum deflection at the roof is

$$\Delta = (.005)(\text{building height}) = (.005)(85)(12) = 5.1"$$

From equation 53, $C_F = ZICS$. However, both Z and I are equal to one. Assuming S also is equal to one, SEAOC equation 1-2 can be used to relate C and T. Then,

$$C_F = C = \frac{1}{15\sqrt{T}}$$

Or,
$$T = .25\sqrt{(5.1)(15)}\sqrt{T}$$

Solving for T yields T = 2.84 seconds

The only assumption made was that S was one. SEAOC equation 1-4A should be used to check that assumption.

$$T/T_S = 2.8/1 = 2.8$$

$$S = 1.2 + .6(2.8) - .3(2.8)^2 = .53$$

Since the SEAOC code limits S to one or greater, the assumption was valid.

14. Other Forces on the Structure

A. PΔ Effect

The column members in a structure are loaded in compression by the dead and live loads (P). When the structure is acted upon by seismic forces, the moment produced by the now eccentric vertical forces will add compressive stresses to the vertical columns. The magnitude of such an overturning moment is written as PΔ.

If the moment were to increase faster than the restoring force from the frame stiffness, the frame would be unstable. Such instability generally is considered only for frames in non-seismic areas which are built to resist only vertical loads. If such a frame is loaded vertically with a constant force, the frame will eventually buckle out from under the load. Since the load is constant (not transient like a seismic load), the (P/A) stress combined with the PΔ stress will cause the column members to fail. Protection against such failure can be provided by X-bracing in the walls or thick shear walls.

Frames designed to withstand large lateral (seismic) loads experience little problem due to PΔ instability.

<u>Figure 29</u>

<u>PΔ Effect</u>

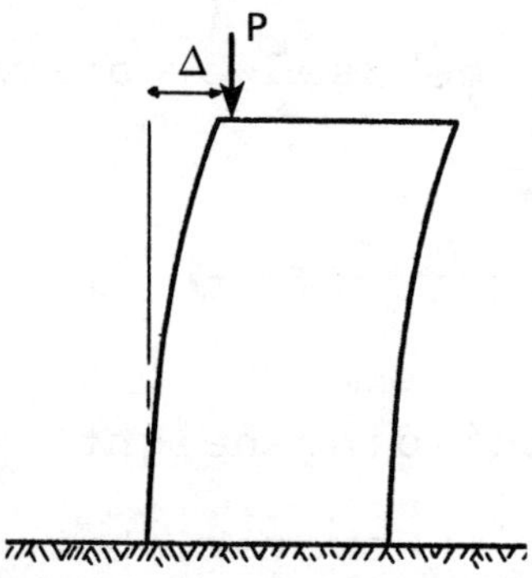

B. Torsional Shear Stress

If the structure's center of mass does not coincide with its center of rigidity, the center of mass will twist about the center of rigidity when the structure is accelerated. This twisting will induce a torsional shear stress which is proportional to the distance from the center of rigidity.

The distance from the center of mass to the center of rigidity is known as the eccentricity. The SEAOC code require that all buildings be designed for at least 5% eccentricity (based on the maximum horizontal building dimension), even if their centers of mass and rigidity coincide. This is known as 'accidental eccentricity.' Of course, if the actual eccentricity is greater than 5%, the actual eccentricity should be used.

Negative torsional shear stress is to be neglected. Thus, the effect of specifying a 5% eccentricity is to slightly overdesign the structure. Even though it is assumed that only walls parallel to the seismic force resist the base shear, all walls resist the torsional shear.

Example 8

Calculate the maximum torsional shear stress induced in the column when a spectral acceleration of .3 gravities in the x direction is experienced.

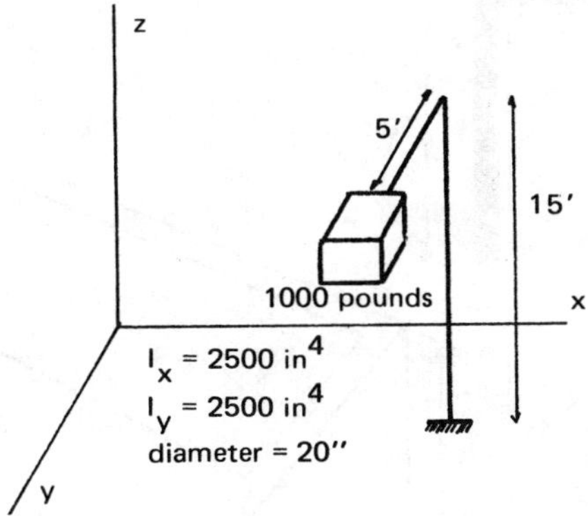

The inertial force is

$$F = ma = (\frac{1000}{32.2})(.3)(32.2) = 300 \text{ pounds}$$

The applied moment (torque) is

$$M = Fe = (300)(5)(12) = 18,000 \text{ inch-lbs}$$

The polar moment of inertia is

$$J = I_x + I_y = 2500 + 2500 = 5000 \text{ in}^4$$

The maximum torsional shear stress is

$$v = \frac{Mr}{J} = \frac{(18000)(10)}{5000} = 36 \text{ psi}$$

Example 9

The 4-wall building shown below experiences a seismic force from the East equivalent to .2 gravities. There are no openings in any of the walls. Each wall is 8" thick and is constructed of reinforced concrete. The roof consists of a 6" slab indicated by dotted lines on the plan view. What are the torsional shears in the walls?

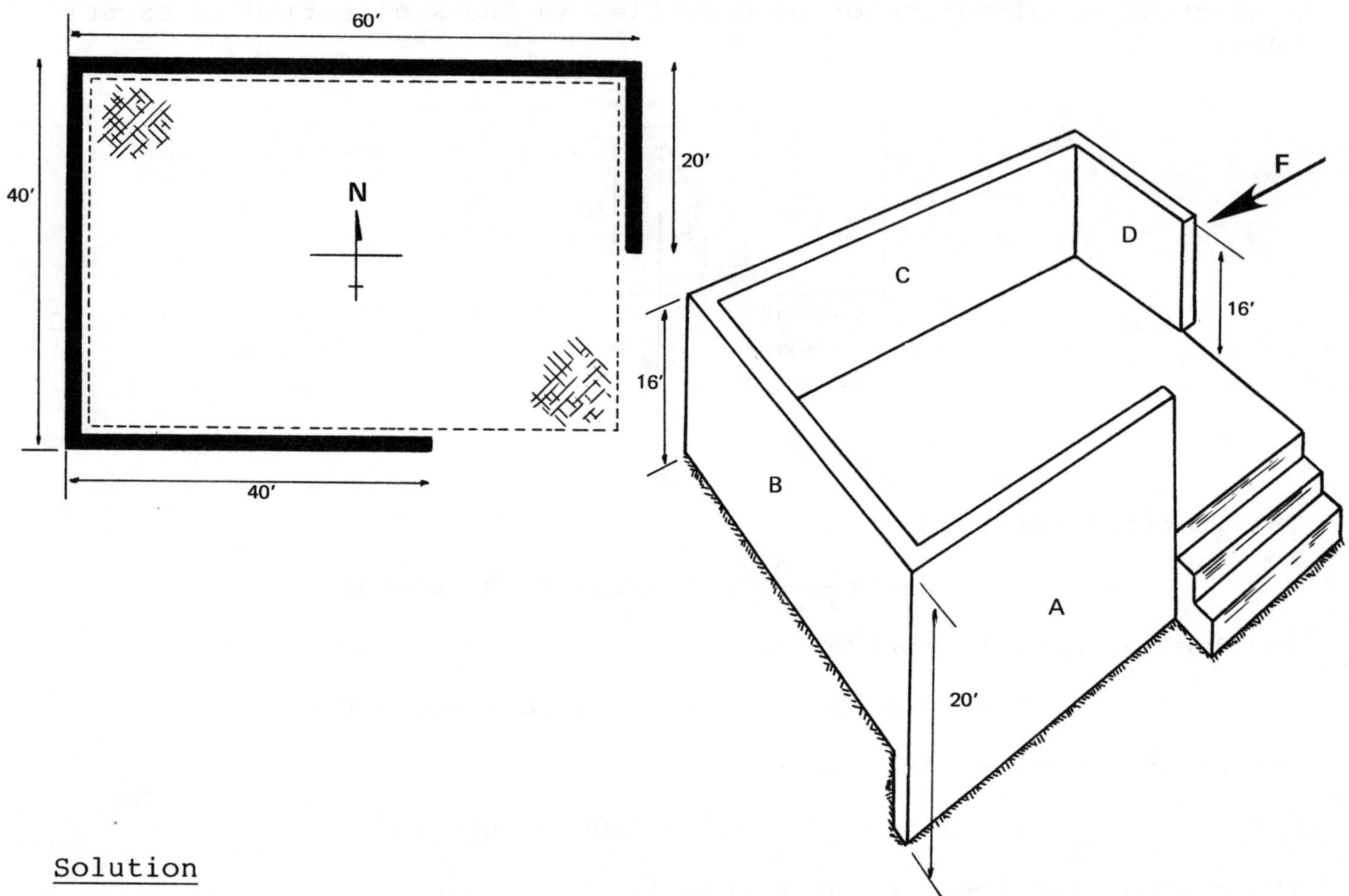

Solution

<u>step 1</u>:   Locate the center of gravity of the structure.

The density of concrete is 150 pcf. The weight of wall A is

$$(8/12)(40)(20)(150) = 80,000 \text{ lb}$$

The center of gravity of the wall (in the plan view) is at (20,0). Similarly, the weights and centers of gravity for all walls are listed below.

element	W/1000	x	y	xW/1000	yW/1000
A	80	20	0	1600	0
B	64	0	20	0	1280
C	96	30	40	2880	3840
D	32	60	30	1920	960
roof	180	30	20	5400	3600
totals	452			11800	9680

So, the center of gravity for the building is at

$$\overline{x}_C = (11{,}800/452) = 26.1$$

$$\overline{y}_C = (9680/452) = 21.4$$

step 2: Locate the center of rigidity.

The method of calculating rigidities of fixed piers is covered in section 18 of this document.(The table on page 95 was used.)

| | Rigidities | |
element	E–W	N–S
A	6.15	0
B	0	7.91
C	12.05	0
D	0	3.43

Now treat the rigidities like weights or forces, and find the centroid of the rigidities (the 'center of rigidity'). The weighting factors are the wall distances from the axes.

$$\overline{y}_R = \frac{(0)(6.15) + (40)(12.05)}{6.15 + 12.05} = 26.5$$

$$\overline{x}_R = \frac{(0)(7.91) + (60)(3.43)}{7.91 + 3.43} = 18.1$$

step 3:   Calculate the eccentricity

$$e = 26.5 - 21.4 = 5.1$$

The SEAOC code specifies a minimum accidental eccentricity of 5% of the maximum building dimension.   Since $(.05)(60) = 3.0$, the 5.1 governs.

step 4:   Calculate the torsional moment.

$$M = Ve = \left(\frac{452{,}000}{32.2}\right)(.2)(32.2)(5.1) = 4.6 \text{ EE5 ft-lb}$$

step 5:   Calculate the torsional resisting force in each wall. Use the following tabular approach. R is the rigidity; d is the perpendicular distance from the wall to the center of rigidity.

wall	R	d	Rd	Rd^2	$M(Rd)/\sum(Rd^2)$
A	6.15	26.5	163.0	4318.8	4956
B	7.91	18.1	143.2	2591.4	4354
C	12.05	13.5	162.7	2196.1	4947
D	3.43	41.9	143.7	6021.7	4370
total				15128.0	

step 6:   Determine which walls receive positive torsional shears and which receive negative torsional shears. This is usually done by inspection.

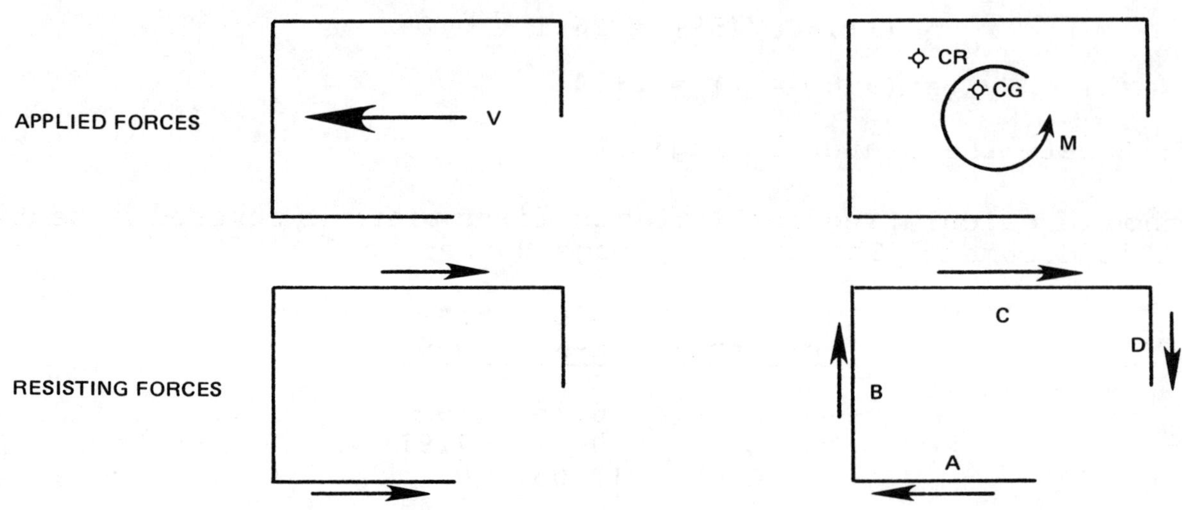

Wall A has a negative torsional shear since the torsional shear opposes the force resisting the base shear. The SEAOC code calls for neglecting negative torsional shears when designing wall A.

step 7:   Calculate the base shears by the procedure in section 18.

C. Overturning Moment

The summation of all moments due to the distributed lateral forces is known as the overturning moment (OTM). If this OTM is great enough, it can overcome the compressive stress in outer columns caused by the dead and live loads. Thus, footings and concrete walls could be placed in a state of tension. The OTM will also increase the compressive stress in outer columns (on the opposite side of the building). Such an increase must be countered by increased thickness of shear walls and extra steel reinforcement in concrete columns.

Two OTM's should be calculated. The first is the sum of all moments taken about the base. This moment should be used to size footings and to design the primary outer columns. The second OTM is the sum of moments taken about a floor level (involving only lateral forces above that floor). This second OTM should be used to design the shear walls and other supporting structures at that floor.

15. Drift Control

Drift is another name for deflection or magnitude of oscillation. Drift needs to be controlled for two major reasons. Excessive drift at the upper stories of a building has a negative psychological effect on occupants. However, the major incentive for controlling excessive drift is to ensure structural integrity. Excessive drift is associated with extreme bending moments and plastic failure. Furthermore, excessive drift is also associated with non-structural damage, such as failure of partitions, windows, ceilings, and elevator shafts.

There are three components of drift. These are:

1.   bending of columns due to flexure and shear stresses in the columns

2.   joint rotation due to stresses in the joints and girders

3.   bending of the frame as a whole unit due to axial stresses in the frame

The first two components together are referred to as shear drift. The third component is known as chord drift.

Figure 30

Shear and Chord Drift

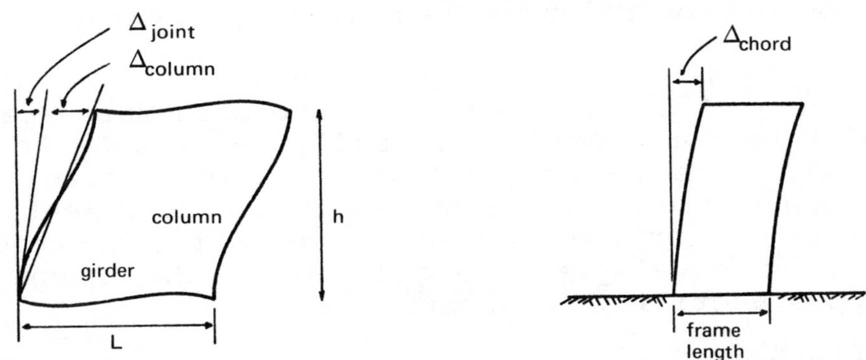

The SEAOC code limits the story drift to (.005) times the story height. Thus, for buildings with uniform plan views and constant story heights, the maximum lateral drift at the upper story is (.005) times the building height. This maximum value is to be compared against the theoretical drift (calculated by structural theory) multiplied by the greater of 1.0 and (1.0/K). (The use of the code will generally result in an actual limitation of total drift of about .5% for moderate earthquakes. However, a total drift of 1.5% can be expected for severe earthquakes.)

Although the calculation of drift is not covered in this section, some generalizations about the relationship between drift and other variables are possible. These variables are covered in table 8.

Table 8

Drift Versus Other Variables

Variable	Column Drift	Girder Drift	Joint Drift	Chord Drift
story height	$(height)^3$	$(height)^2$	$(height)^2$	
building height				$(height)^3$
girder length		$(length)^1$		$(length)^3$
column depth	$(depth)^2$		$(depth)^{-1}$	
girder depth		$(depth)^2$	$(depth)^{-1}$	
column height	$(height)^1$		$(height)^2$	
shear load				$(load)^1$
frame length				$(length)^{-2}$

16. Seismic Provisions for Concrete Design

A. Introduction

One of the strong points of reinforced concrete structures is that almost all connections can be considered rigid. This is due to the construction, which is either monolithic or modular with cast-in-place connections. Thus, the entire frame responds to vertical and lateral loads even if the load is applied to only one part. If there is a weak part in the structure, some of the load at that weak point will be carried by stronger parts of the structure.

All structural elements (including shear walls and framework) are usually connected at every floor level by reinforced slabs. Since slabs have high rigidity, they transmit all lateral loads to framework and shear walls. The amount of the lateral load carried by each resisting element will be in proportion to its rigidity, as explained later in this document.

Since all members of a reinforced concrete structure are fairly rigid, the drift will be correspondingly small. Although the ductilty of a concrete framework is not as great as it is for a steel framework, the ductility is more than adequate to meet the requirements of earthquake codes. However, attention must be given to the arrangement of reinforcement and to development lengths since the concrete itself has relatively poor ductility in shear.

The primary disadvantage of concrete is its weight. This weight increases the seismic forces. Some advantage can be gained in the use of lightweight structural concrete.

B. The Concrete Code

The following is a synopsis of some of the more important provisions of Appendix A of ACI 318. This synopsis is not complete, and significant aspects of the code have been simplified. This material is included to illustrate some of the more important detailing requirements to ensure adequate seismic performance of ductile frames. (Seismic provisions of the UBC for reinforced concrete are similar but do have many differences.)

Design Strengths

(a) The ultimate compressive strength of concrete shall not be less than 3000 psi.

(b) The yield strength of the reinforcing steel shall not exceed 60,000 psi.

(c) Higher grades of steel shall not be substituted for lower grades.

Flexural Members

(a) The reinforcement ratio shall not be greater than 50% of the ratio that would produce balanced strain. However, the reinforcement ratio shall not be less than $200/f_y$.

(b)   Top and bottom reinforcement shall consist of not fewer than 2 bars throughout the length of the member.

(c)   Flexural members framing into opposite sides of a column shall have continuous reinforcement between them.

(d)   If it is not possible to continue reinforcement through a column (because of variations in the member sections or because there is no member on the opposite side) member reinforcement shall terminate within the column with a standard 90 degree hook.

(e)   Development length shall not be less than (2/3) the standard development length, nor less than 16 inches.

(f)   Web reinforcement (using #3 bars or larger with (d/2) spacing or less) shall be use to resist both gravity and lateral loads.

Splices

(a)   The minimum lap splice length is 24 bar diameters, but not less than 12 inches for flexural members, and 30 bar diameters but not less than 16 inches for columns.

(b)   At least 2 stirrup ties are to be used at each lap splice.

(c)   Welded splices are not to be located closer than a distance (d) from a plastic hinge.

Columns Subjected to Axial and Flexural Loading

(a)   The column reinforcement shall be at least .01 but less than .06 times the gross column area.

(b)   In general, the moment capacity of a column must exceed the sum of moment capacities of all flexural members tributory to it at a particular point.

(c)   If the maximum factored axial earthquake load exceeds $(.4)\phi$ times the calculated ultimate strength of the column, special confinement reinforcement shall be used above and below beam-column connections.

(d)   Special confinement reinforcement is also required for columns which support walls and stiff partitions that do not continue from story to story.

(e)   Special transverse reinforcement is required in columns to carry shear occurring when joints develop their maximum moment carrying capacity.

Shear Walls

(a)   Shear walls shall be designed to transmit their loads to the foundations.

(b)   The area of both horizontal and vertical reinforcement shall not be less than (.0025) times the gross wall area.

(c)   Shear walls shall be constructed with vertical boundary elements to support the total vertical load from factored and lateral loads. Such vertical boundary elements shall have full-length confinement.

Two important aspects of the above provisions bear special mentioning. The first pertains to the use of additional confinement in columns and shear wall boundary elements. This confinement is used to resist shearing forces, to prevent the main bars from buckling, and to increase the column ductility. All main bars should be wrapped. In the case of columns with large sectional areas, extra hoops should be arranged inside of the main bars to prevent them from bucklig.

<u>Figure 31</u>

Concrete Confinement

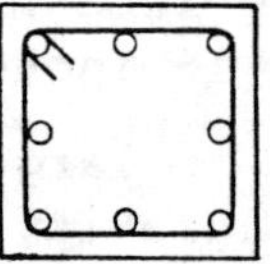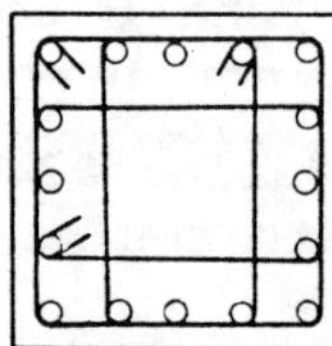

Since columns always have some confinement (ties or spirals), the confinement called for in the code is additional. This is illustrated in figure 32.

<u>Figure 32</u>

Confinement Spacing

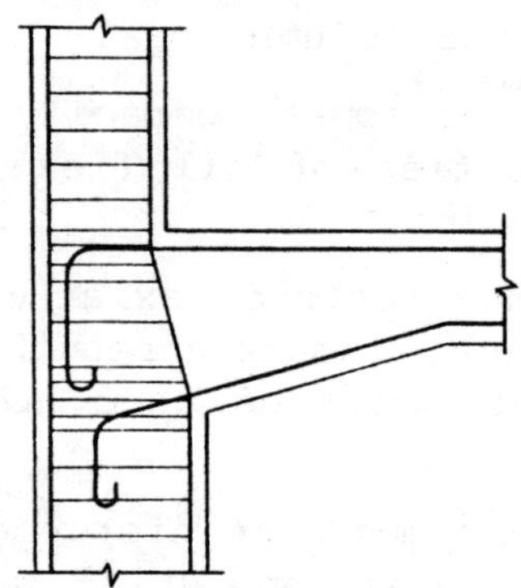

The other important provision pertains to the use of shear wall boundary elements. This essentially means that shear walls shall be framed on their vertical sides by column members proportioned to carry the entire vertical loading on the wall. This is illustrated in figure 33.

Figure 33

Shear Wall Boundary Elements

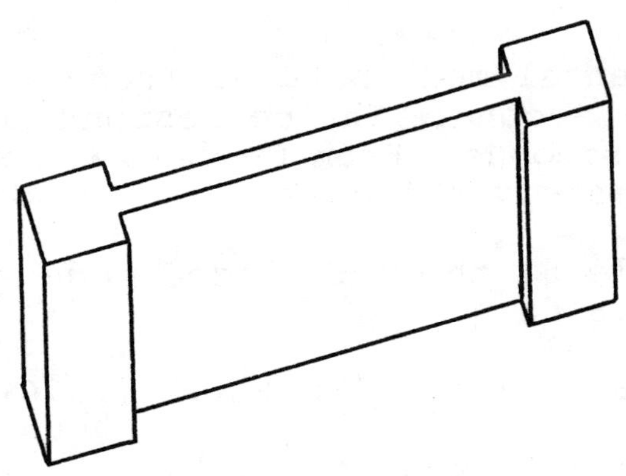

17. Distributing Base Shear to Stories

A. Mode Shape Analysis

If the mode shape is known for a MDF system, the following procedure may be used to find the story shears. (It is also necessary to know the inter-story spring constants, the natural frequency, and the damping coefficient.)

step 1:   Given the natural frequency and damping, go to the response spectra and find S_d. If necessary, use any of the relationships between S_v, S_a, and ω to find S_d from spectral acceleration or velocity.

step 2:   Calculate the modal participation factor. (ϕ is the normalized mode shape factor.)

$$\Gamma = m_1\phi_1 + m_2\phi_2 + m_3\phi_3 + \cdots + m_n\phi_n \qquad 56$$

step 3:   Multiply the modal participation factor by S_d and each mode shape factor. This will give the actual deflection at each story.

step 4:   Calculate the relative displacements between floors to obtain the displacement resisted by the springs.

step 5:   Multiply the relative displacements by the spring constants.

<u>Example 9</u>

Determine the base shears for the MDF system evaluated in example 4. Consider only the fundamental mode. Use the El Centro response spectra with 5% damping.

<u>step 1</u>: The fundamental mode natural frequency was found to be 5.18 radians per second. The corresponding period (from equation 15) is 1.2 seconds. From figure 22, the spectral displacement is approximately 4 inches.

<u>step 2</u>: From example 5, the normalized mode shape factors are .408, .707, and .816.

$$\Gamma = (1)(.408) + (1)(.707) + (.5)(.816) = 1.523$$

<u>step 3</u>:
$$x_1 = (1.523)(4)(.408) = 2.49"$$
$$x_2 = (1.523)(4)(.707) = 4.31"$$
$$x_3 = (1.523)(4)(.816) = 4.97"$$

<u>step 4</u>:
$$x_{1-ground} = 2.49"$$
$$x_{2-1} = 4.31 - 2.49 = 1.82"$$
$$x_{3-2} = 4.97 - 4.31 = .66"$$

<u>step 5</u>: Since the spring constants are each 100, the cumulative story shears are

$$V_1 = 249$$
$$V_2 = 182$$
$$V_3 = 66$$

◆

B. Participation Factor Method

The analytical method just presented is not actually used in practice because the stiffnesses are not known in advance. Even if all 'spring' members were known, some of them would be redundant. It is also quite unlikely that the force-deflection curve would be linear.

The following simple method may be used to calculate the story shears in such a case. Of course, the mode shape factors (ϕ_{ij}) must still be known.

Let F_{ij} be the story shear for the ith story while vibrating in the jth mode. Similarly, let ϕ_{ij} be the mode shape factor for the ith story vibrating in the jth mode. Let W_i be the weight of the ith story. Then,

$$F_{ij} = (W_i\phi_{ij}S_a)\frac{\sum W_i\phi_{ij}}{\sum W_i\phi_{ij}^2} = W_i\phi_{ij}S_a\Gamma_j \qquad 57$$

This method also permits the use of unnormalized values of ϕ_{ij}. Equation 58 gives the modal participation factor.

$$\Gamma_j = \frac{\sum W_i \phi_{ij}}{\sum W_i \phi_{ij}^2} \qquad\qquad 58$$

Example 10

Solve example 9 using equation 57.

Solution

From figure 20, the spectral acceleration is .28 gravities. This corresponds to 9.0 ft/sec^2 and 108 in/sec^2.

Assuming the ϕ_{ij} are unnormalized (which they are not), the modal participation factor is

$$\Gamma = \frac{(1)(.408) + (1)(.707) + (.5)(.816)}{(1)(.408)^2 + (1)(.707)^2 + (.5)(.816)^2}$$

$$= \frac{1.523}{1.0} = 1.523$$

From equation 57, the story shears are

$$F_3 = (.5)(.816)(108)(1.523) = 67.1$$

$$F_2 = (1)(.707)(108)(1.523) = 116.3$$

$$F_1 = (1)(.408)(108)(1.523) = 67.1$$

The cumulative story shears are

$$V_3 = 67.1$$

$$V_2 = 67.1 + 116.3 = 183.4$$

$$V_1 = 67.1 + 116.3 + 67.1 = 250.5$$

C. SEAOC Code Method

The SEAOC code contains a specific method of distributing the base shear calculated from equation 55 to each story. This method of distribution is sometimes called 'throwing weight to the top' since the greatest inertial shear load is applied to the top story. The distribution shape is illustrated in figure 34.

Figure 34

Base Shear Distribution

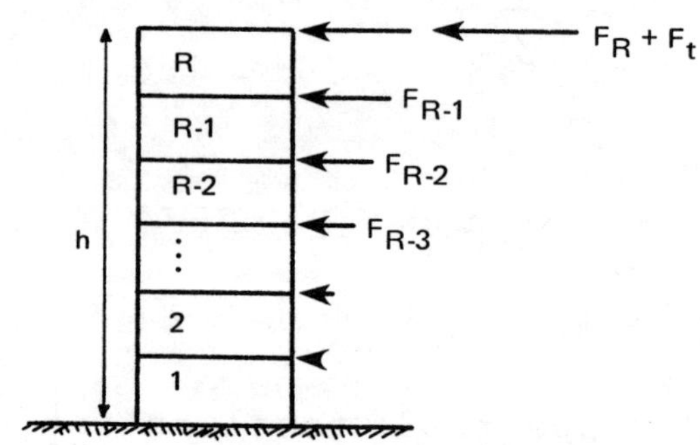

The force applied to the top is comprised of two components. The first component is given by equation 59.

$$F_t = (.07)TV \qquad\qquad 59$$

F_t is limited to $(.25)V$, and it may be taken as zero if T is less than .7 seconds.

The second component, F_R, and all forces applied to lower levels are calculated recursively from equation 60. For the ith story, the force is

$$F_i = \frac{(V-F_t)W_i h_i}{\Sigma W_i h_i} \qquad\qquad 60$$

The summation is taken from 1 to n for each calculated force.

In equation 60, W_i is the weight of the ith story. Similarly, h_i is the height of the story above the ground as measured to the floor of that story.

Example 11

A 5-story building is to be constructed with a ductile moment-resisting frame. The story height is 12 feet. Its base shear is calculated as 160 kips. The roof weight is 700 kips. All floors weigh 800 kips. What is the distribution of the lateral force to each story if the natural period is 2.8 seconds?

Solution

The top force, F_t, is

$$F_t = (.07)(2.8)(160) = 31.4 \text{ kips}$$

A table is the easiest way to set up the data for finding values of F_R.

Level	h_i	W_i	h_iW_i	$h_iW_i/\Sigma h_iW_i$
R	60	700	42,000	.304
5	48	800	38,400	.278
4	36	800	28,800	.209
3	24	800	19,200	.139
2	12	800	9,600	.070
1	0	0	0	0
		total	138,000	1.000

The value of F_R is $(160 - 31.4)(.304) = (128.6)(.304) = 39.1$

Similarly,

$$F_5 = (128.6)(.278) = 35.7$$

$$F_4 = (128.6)(.209) = 26.9$$

$$F_3 = (128.6)(.139) = 17.9$$

$$F_2 = (128.6)(.070) = 9.0$$

◆

18. Distributing Story Shears: Analysis

The lateral forces calculated in the previous section were assumed to be applied at the ceiling height of a story. The ceiling does not actually resist the lateral load applied at that level, but it does distribute the force among the shear walls and columns. This is illustrated in figure 35.

<u>Figure 35</u>

<u>Distribution of Lateral Forces to Shear Walls</u>

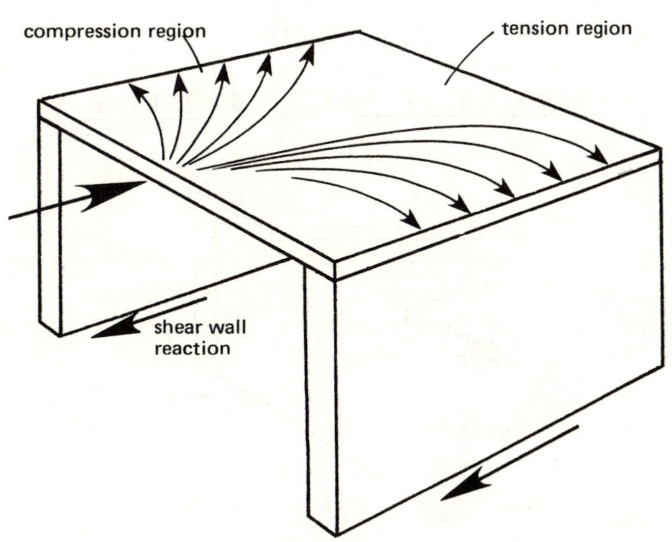

Ceilings and floors that transmit lateral forces to the resisting elements (shear walls and columns) are known as horizontal diaphragms. Diaphragms may be constructed of concrete slabs or heavy metal decking (rigid diaphragms). Or, they may consist of wood sheathing or light metal decking (flexible diaphragms). Rigid diaphragms are assumed not to deflect due to the application of a lateral load.

Another major difference between flexible and rigid diaphragms pertains
to their ability to transmit torsion. Rigid diaphragms are considered
capable of transmitting torsion. Flexible diaphragms are not.

A. Rigid Diaphragms

The action of rigid diaphragms is governed by the following two rules.

> (1)   Any deflection that occurs at one point in the structure
> also occurs at all other points in the structure.

> (2)   Rigid diaphragms distribute the lateral load to the
> resisting elements in proportion to the rigidities of the
> resisting elements.

Rigidity is defined as the reciprocal of deflection. The deflection
experienced by a resisting element is caused by the application of both
moments and shears.

Consider the shear wall shown in figure 36. This shear wall is fixed at
both the top and bottom. Deflection is accompanied by double curvature,
since both the top and bottom of the wall must remain perpendicular. The
deflection (the reciprocal of rigidity) is given by equation 61. For
masonry, E is typically taken as 1,000,000 psi and G is taken as (.4)E.
For steel, G = (.38)E. For concrete, E = 3,000,000 psi and G = (.4)E.

$$\Delta = \frac{Fh^3}{12EI} + \frac{(1.2)Fh}{AG} \qquad\qquad 61$$

$$A = td \qquad\qquad 62$$

$$I = \frac{td^3}{12} \qquad\qquad 63$$

Figure 36

Fixed Shear Wall Deflection

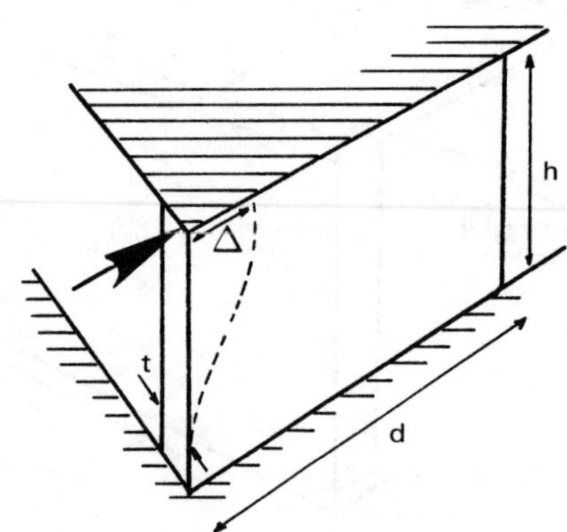

For a cantilever wall which is allowed to bend in simple curvature, equation 64 may be used to calculate the deflection.

$$\Delta = \frac{Fh^3}{3EI} + \frac{(1.2)Fh}{AG} \qquad\qquad 64$$

Figure 37

Cantilever Wall Deflection

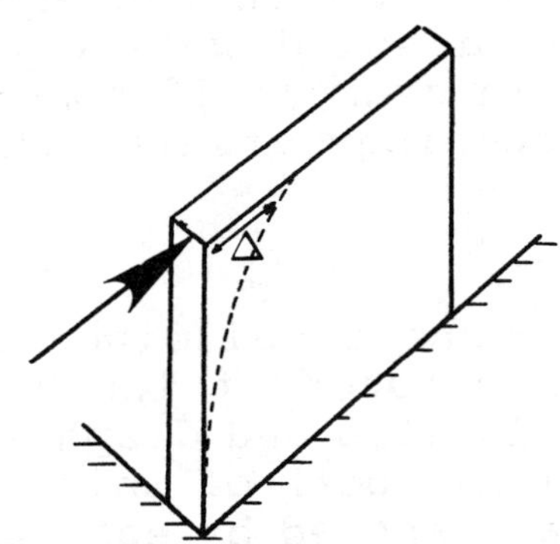

Since the shear is distributed to the walls and piers in proportion to their rigidities, the actual values of the deflections and rigidities are not needed. Only the ratios of individual rigidities to the total rigidity (the 'distribution factors') are needed. Therefore, F and t can be given arbitrary values of 100,000 and 1.0 respectively. The rigidities of the fixed and cantilever walls are given below. Equations 65 and 66 assume E = 1,000,000 psi.

$$R_{fixed} = 1/[.1(h/d)^3 + .3(h/d)] \qquad\qquad 65$$

$$R_{cantilever} = 1/[.4(h/d)^3 + .3(h/d)] \qquad\qquad 66$$

The following rules should be observed:

(a) If the wall extends above the roof height, only the distance from the ground to the roof is to be used in calculating the deflection.

(b) The shear stress should not be allowed to exceed the limits given in the UBC for walls.

(c) Positive torsional shear stress should be added. If the centers of rigidity and gravity coincide, a torsional shear stress due to a 5% accidental eccentricity should be added.

(d) Normally, the rigidities of the transverse walls are neglected. This is called 'omitting the weak wall.'

(e) In multi-story buildings, the total force to be distributed also includes all story loads above the level whose piers are being evaluated or designed.

(f)   These procedures apply equally well to shear wall and framed construction as long as the geometry of the resisting elements is known. In the case of vertical frame resisting elements, the rigidity due to shear stress is often neglected since it is so small.

(g)   As shown in example 12, walls with openings have some special considerations.

(h)   The rigidity of a wall with openings should be compared against the rigidity of a solid wall of the same dimensions. The procedure used in example 12 sometimes results in rigidities of walls with openings greater than of solid walls.

Example 12

A one-story reinforced concrete building is constructed with a rigid diaphragm. A total seismic force of 120,000 pounds is applied to the lateral walls. The lateral walls and diaphragm transmit the load to the 9" thick shear walls (A and B), both of which are shown in elevation view below. Determine the shear carried by each wall. Neglect torsional and overturning effects.

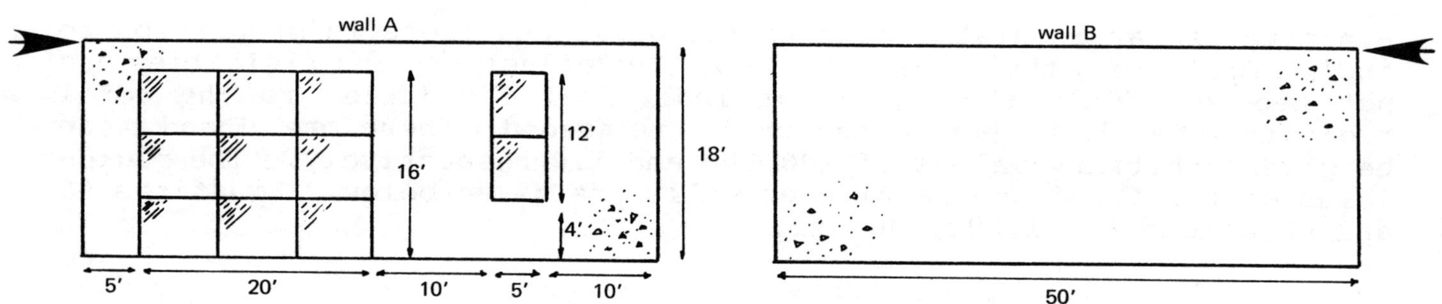

Solution

It is necessary to divide wall A into piers and beams. A pier is a vertical portion of the wall whose height is taken as the smaller of the heights of the openings on either side of it. Beams are the horizontal portions left over after the piers have been located. For wall A, P1, P2, and P3 are piers. B1 and B2 are beams.

For pier P1, $h/d = 16/5 = 3.2$

$$R_{P1} = 1/[.1(3.2)^3 + .3(3.2)] = .236$$

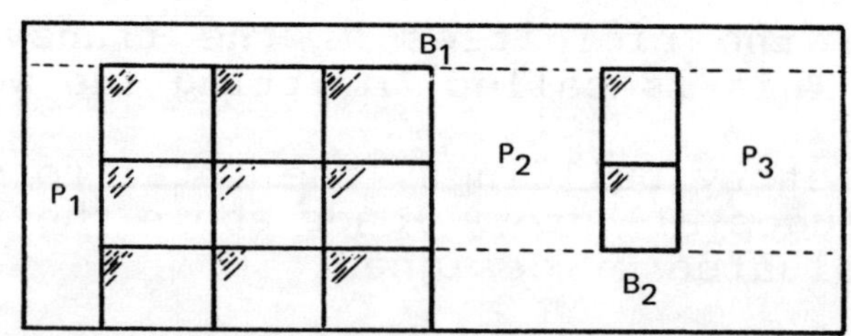

Similarly, the rigidities of the other piers and beams are

	h	d	h/d	R	1/R
P2	12	10	1.2	1.877	
P3	12	10	1.2	1.877	
B1	2	50	.04		.0120
B2	4	25	.16		.0484

The total rigidity of wall A is

$$R_A = \cfrac{1}{\cfrac{1}{.236 + 1.877 + 1.877} + .0120 + .0484} = 3.215$$

The total rigidity of wall B is found to be 8.876.

The total base shear is distributed to the walls in proportion to the rigidities.

$$V_A = (120,000)\left(\frac{3.215}{3.215 + 8.876}\right) = 31,908 \text{ lb}$$

$$V_B = (120,000)\left(\frac{8.876}{3.215 + 8.876}\right) = 88,092 \text{ lb}$$

◆

B. Flexible Diaphragms

Flexible diaphragms are not capable of distributing seismic forces in proportion to the rigidities of the resisting elements. Rather, the assumption is made that the forces are distributed to the resisting elements in proportion to the tributary areas of the diaphgragm. It is assumed that drag struts (also known as 'collectors' or 'braces') are used to transmit diaphragm loads to the shear walls at points of discontinuity in the plan. Example 13 illustrates the distribution method.

Example 13

The plan view of an L-shaped building is shown below. Determine the tributary areas for earthquakes in both the N-S and E-W directions.

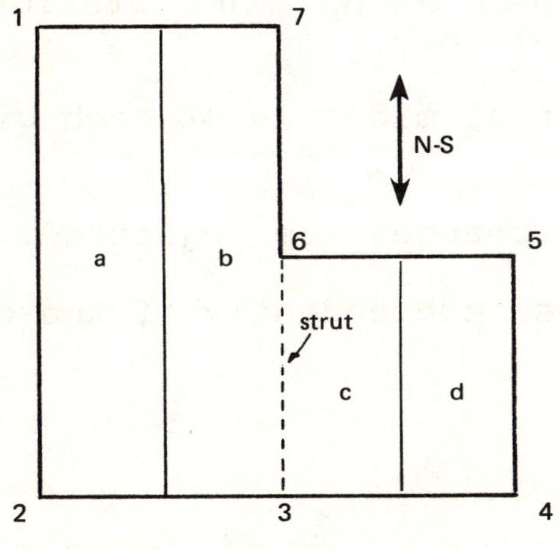

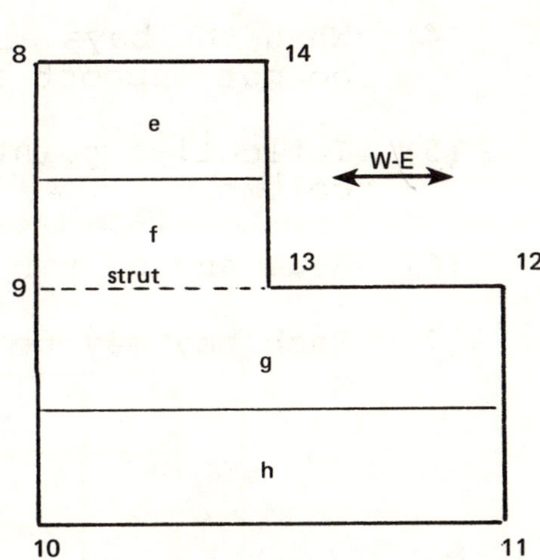

<u>Solution</u>

wall section	tributary areas
N-S earthquake	
1-2	a
4-5	d
6-7	b and c
E-W earthquake	
8-14	e
10-11	h
12-13	f and g

◆

19. Distributing Story Shears: Design

Section 18 dealt with distributing story shears in structures whose resisting elements were already designed. If the shape, area, and moment of inertia of the resisting elements are not known, some assumptions must be made about the proportion of the base shear that each element will carry.

There are a number of approximate methods that distribute the story shears to the resisting elements. These approximate methods eliminate the need to use indeterminate solution methods (such as the moment distribution method) to evaluate the forces taken by each resisting element.

A. The Portal Method

In the portal method, the following assumptions are used:

(1)   All interior columns carry the same shear.

(2)   Exterior columns carry half of the shear carried by interior columns.

(3)   If the bays (length between the columns) are not the same length, the shear may be distributed to the columns in proportion to the length between them.

(4)   When the bays are of equal length, the interior columns do not support any axial load.

(5)   Inflection points occur at mid-span of each girder and column.

(6)   Beam and column length changes are neglected.

(7)   Each bay may be treated independently of the others.

Figure 38

Portal Method Frame Deflection

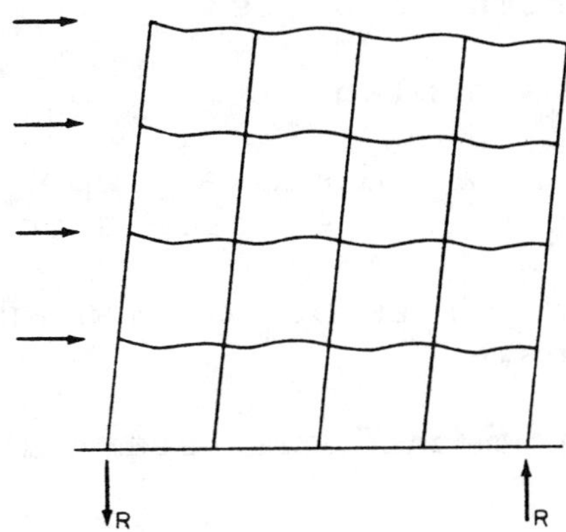

Example 14

Consider the frame shown below. The sum of story shears for layer A and above is 110 kips. Find the forces on members 2-3 and 2-7. Use the portal method of analysis.

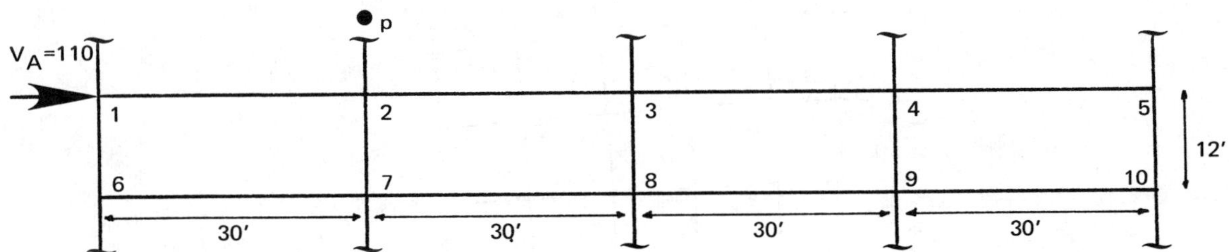

Solution

Since columns 1-6 and 5-16 carry only 1/2 the loads carried by the interior columns, the total shear carried by all five columns is

$$2\left(\tfrac{1}{2}V_c\right) + 3(V_c) = 4(V_c)$$

So, $V_c = 110/4 = 27.5$ kips

Taking the free body of joint 2 and summing moments abut point P will give the girder shears.

$$\Sigma M_p = (2)(27.5)(6) - (2)(V_g)(15) = 0$$

Or, $V_g = 11$ kips

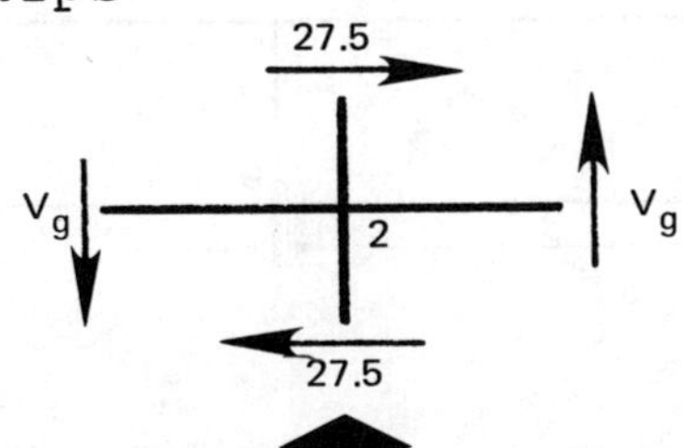

B. The Cantilever Method

Unlike the portal method which treats each bay independently of the others, the cantilever method assumes the entire floor works together as a unit. The cantilever method is preferred for buildings taller than 25 stories. The primary assumptions are:

    (1)   Floors remain plane.

    (2)   The force in a column is proportional to the distance of the column from the frame's center of gravity.

    (3)   Inflection points of columns and girders occur at their mid-lengths.

Analysis by the cantilever method must start at the roof level and work down.

Figure 39

Cantilever Method Frame Deflection

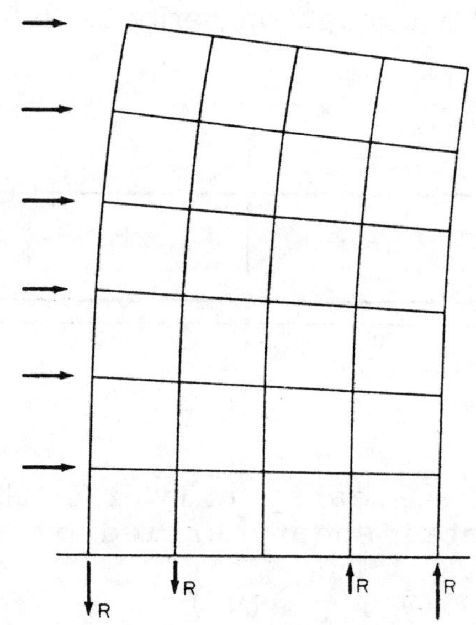

Example 15

Consider the frame shown below. The lateral forces at the roof level total 57 kips. What are the forces in members 2-3 and 2-7? Use the cantilever method.

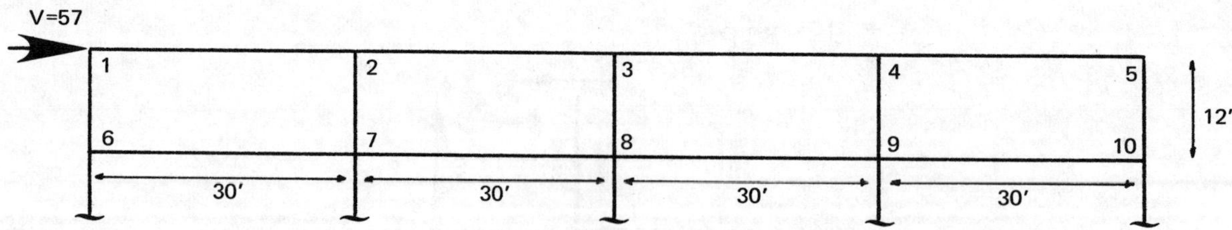

Solution

The center of gravity of this layer is at point P. Columns 1-6 and 2-7 are in tension. Columns 4-9 and 5-10 are in compression. Column 3-8 is not stressed axially. Since the bays are all the same size,

$$T_{1-6} = 2T_{2-7} = 2C_{4-9} = C_{5-10}$$

Summing moments about point P gives

$$\Sigma M = (60)T_{1-6} + (30)T_{2-7} + (30)C_{4-9} + (60)C_{5-10} = (57)(6)$$

Or,
$$T_{1-6} = C_{5-10} = 2.28 \text{ kips}$$

$$T_{2-7} = C_{4-9} = 1.14 \text{ kips}$$

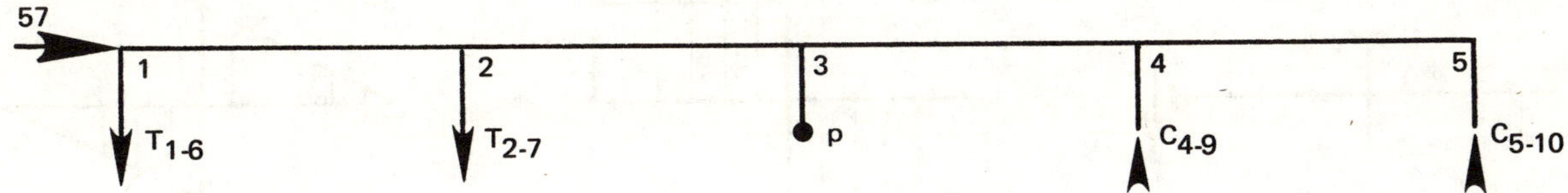

The vertical shear in member 1-2 is equal to the sum of vertical loads to the left of the inflection point for girder 1-2.

$$V_{1-2} = T_{1-6} = 2.28$$

Summing moments about point 1 will give the horizontal shear in member 1-6.

$$\Sigma M = (2.28)(15) - (V_{1-6})(6) = 0$$

$$V_{1-6} = 5.7 \text{ kips}$$

The vertical shear in member 2-3 is $(1.14 + 2.28) = 3.42$ kips. Summing moments about point 2 will give the horizontal shear in member 2-7.

$$\Sigma M = (2.28)(15) + (3.42)(15) - (V_{2-7})(6) = 0$$

$$V_{2-7} = 14.25 \text{ kips}$$

20. Design of Frame Members

Once the shears are known for all columns and girders, the sizing of these members can be done with standard methods.

A. Traditional Sizing

The forces on each member can be used to draw the shear and moment diagrams. Selection of structural shapes should be based on the maximum moments and shears encountered.

Figure 40

Shear and Moment Diagrams

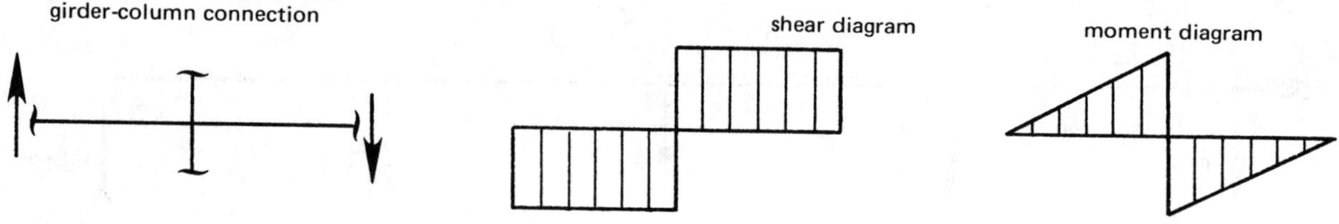

B. Sizing Based on Drift

Since the SEAOC code limits drift to (.005) times the story height, this limitation may be used to size the members. The actual drift due to the application of a column shear load is

$$\Delta = (.005)h = \frac{V_c(h^2)}{12E}\left(\frac{h}{I_c} + \frac{L}{I_g}\right) \qquad\qquad 67$$

There are many possible column and girder combinations that will satisfy this drift criterion.

Figure 41

Portal Drift

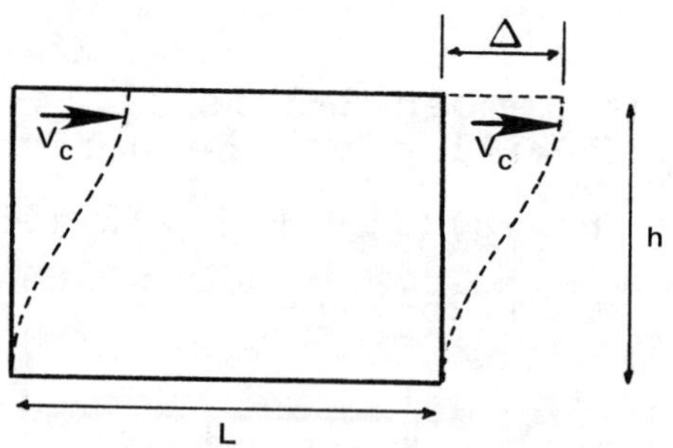

C. The Spurr Method

Although the cantilever method assumes that the floors will remain plane, that method does nothing to ensure the planeness of the resulting design. If the height-to-width ratio is less than 4, the floors will indeed remain fairly plane. However, when the height-to-width ratio exceeds 4, exterior column elongation and shortening cause noticeable deformation in the floors.

The Spurr Method uses a definite relationship between the moments of inertia in the girders. Essentially, the Spurr method requires the moments of inertia of the girders to be proportional to the girder shear times the square of the span if the inflection points are to be located at mid-span.

21. Practice Examination Questions

Question 1

A 17-story building has plan dimensions of 80' x 160'. It has a moment resisting space frame with a K factor of 0.67. The wind pressure is a map area of 30 psf per the UBC. You may assume beam and girder framing for your discussion. Spans are not to exceed 30'. The building is located in seismic zone 3.

(a) What are the possible modes of failure due to seismic forces for this structure if the frame is structural steel?

(b) What are the possible modes of failure due to seismic forces if the structural frame is reinforced concrete?

(c) If the frame is reinforced concrete, what considerations must apply if the K = 0.67 factor is to be used? What design method is preferred in this case?

(d) How can ductility be achieved in the concrete frame? What is the most important consideration? What is the aim?

(e) Distinguish between 'stiffness' and 'rigidity' as used in seismic design considerations. Write the equation for the rigidity 'R' in terms of a single wall pier of height 'H', width 'd', and thickness 't'.

Question 2

(a) Define the term 'ductility factor' as it applies to a structural system.

(b) What factors influence the magnitude of the ductility factor? What is the minimum ductility ratio that is recommended?

(c) What is the principal reason for specifying a minimum ductility factor?

(d) Why will theoretical analyses of elastic response of structures subjected to basic ground motions usually overestimate the stresses produced?

(e) Explain the phenomenon known as the 'PΔ' effect.

(f) In the application of the base shear formula V = ZIKSCW, what factor of K should apply if a structure requires a K of 0.80 in one direction and a K of 1.33 in the other direction? What is the basis for your answer?

(g) Explain in general terms how the base shear V is distributed in a horizontal plane to the various resisting elements.

Question 3

(a) Distinguish between a moment resisting space frame and a ductile moment resisting space frame.

(b) What is meant by a 'box system' as related to a structural system?

(c) Explain the term 'T' as it appears in the equation

$$T = \frac{(0.05)h_n}{\sqrt{D}}$$

(d) What is meant by the term 'confined concrete'?

(e) Describe the two connotations of the term 'drift'.

(f) What is the basis of the K factor in the formula V = ZIKSCW?

(g) What K factor is used for structures other than buildings? What is the basis for this value?

(h) What is the recognized dividing line between ordinary shear wall construction and the need for a ductile moment resisting frame as measured in structure height?

(i) How does the 'Portal Method' as used in tall structures treat the effects of column shortening?

Question 4

(a) Using the Richter scale, what is the lowest acknowledged numerical value that would be used to identify a major earthquake?

(b) What is the upper numerical limit on the Richter scale?

(c) What does the Richter scale measure?

(d) The term 'redundancy' is often applied to the structural framing of a modern high-rise building. What does it mean, and what has been the trend in modern building design? Is redundancy increasing or decreasing, and why?

(e) What is the purpose and function of the spiral looped ties used in a concrete column?

(f) Several factors may influence the magnitude of the ductility factor that can be achieved in a structure. Identify at least four of these factors.

(g) What is 'negative torsional shear'? How is it normally treated?

(h) What is the 'Rayleigh method' and where would it be used?

(i) Explain the term 'critical damping'. How is the amount of damping related to the critical damping? What is the practical range of damping factors? To what extent does damping affect the natural period of vibration of a structural frame?

Question 5

The figure below represents an existing power transformer on a pedestal whose structural adequacy for the design seismic environment must be demonstrated.

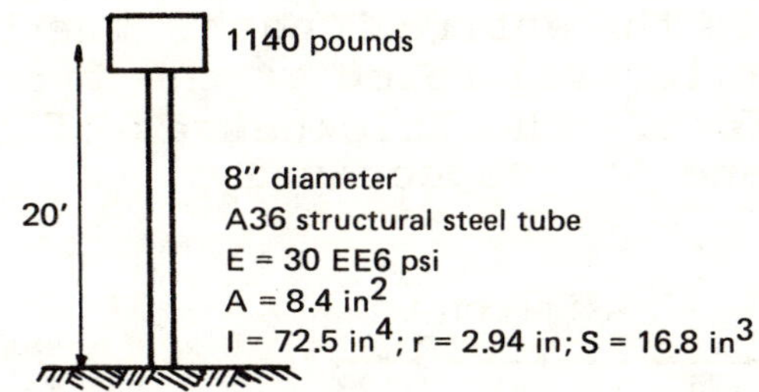

The horizontal component of the environment is defined by the El Centro elastic response spectra. (If the frequency of the system which is excited by the horizontal component exceeds 30 Hz, the response is defined to be equivalent to .5g static load.) The vertical response is defined to be 50% of the horizontal response, and both are assumed to occur simultaneously.

From a test performed on the system, the ratio of successive amplitudes was found to be .882 for free, lateral vibration.

Assume that the shear flexibility of the pedestal, the flexibility of the foundation, the mass of the pedestal, and the rotary inertia of the transformer can be neglected in the following calculations.

(a) Calculate the undamped lateral frequency and period of the system.

(b) Calculate the undamped vertical frequency and period of the system.

(c) How do the above assumptions influence the calculated frequencies?

(d) What is the logarithmic decrement of damping for lateral vibration?

(e) What is the percent of critical damping for lateral vibration?

(f) From the shock spectra determine the lateral response.

(g) Detemine the vertical response.

(h) Determine the maximum base reactions.

(i) Were the assumptions used to calculate the frequencies conservative with respect to the lateral load?

(j) Assume the mode of failure for the structure is yielding of the pedestal due to bending stress. Would you expect a high ductility ratio?

Question 6

The plan views of a two-story concrete structure are shown below. This structure has no openings in the shear resisting walls, although only the east wall covers the entire length. The lateral force at the second floor is 50 kips; the lateral force at the first floor is 260 kips. The story height is 12 feet. The thicknesses of the roof, wall, and floor slabs are 5", 10", and 5" respectively.

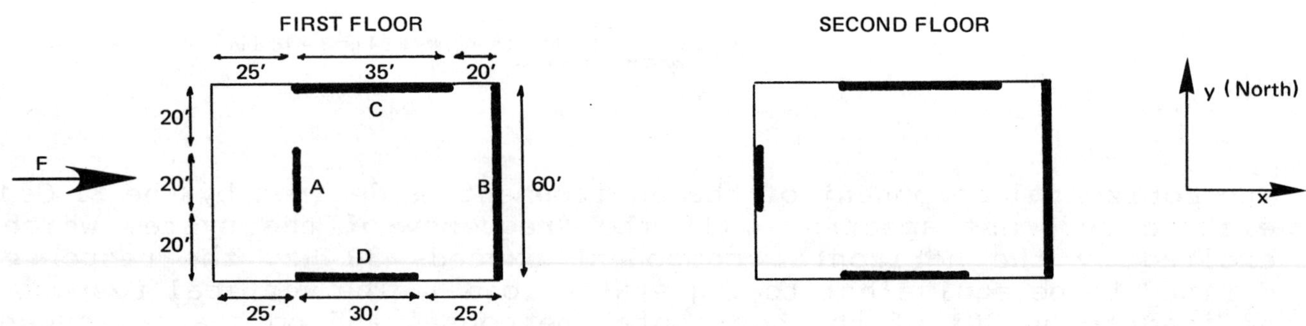

(a) Find the relative rigidities of the second floor walls.

(b) Find the center of rigidity of the second floor walls assuming the center of mass is at the plan center.

(c) How would you solve for the shears in the second floor walls? List the steps. Do not solve for values.

(d) How would you handle positive and negative torsion shears?

(e) How is the first floor effected by offsetting wall 'A' as shown?

Question 7

(a) A building can be built using either a moment resisting steel frame or a moment resisting concrete frame. Which building is more likely to have a smaller natural frequency of vibration?

(b) Which building will have a lower 'C' factor? Explain.

(c) Which building will more likely to have a greater damping effect? Explain.

(d) What is meant by response spectrum?

(e) Draw the response spectrum of an earthquake like the San Fernando earthquake of 1971.

(f) Define what 'confined concrete' is and describe the methods used to obtain confinement.

(g) Describe how a design will treat 'base shear' per the UBC. What factors are involved?

(h) A 9-story building is to be constructed. Draw the seismic force diagram acting on the building.

(i) Draw the shear distribution diagram acting on the same building.

(j) What special details are used in reinforced concrete construction for earthquake protection?

Question 8

The structure shown below is subjected to a lateral loading of .3g. The columns are set in hard rock concrete footings. The reinforced concrete slab is 6" thick and has a finished density of 150 pcf. The columns are rectangular steel (A-36) TS 5x5x(5/16") tubing.

Ignore the dead load of the columns, any live load on the concrete slab, rotation of the slab, and deflection of the columns. Determine if the columns are adequate. Show all calculations.

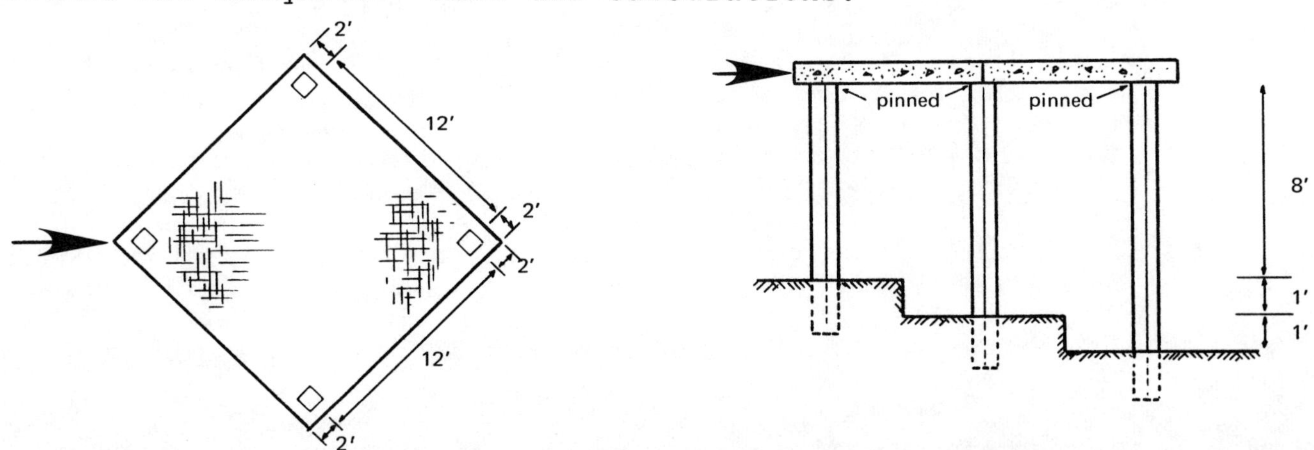

Question 9

(a) What is drift? What limitation on drift is specified by the SEAOC code?

(b) What is critical damping? How does one relate the damping of buildings to their critical damping?

(c) What method is used to distribute a known horizontal shear among the resisting elements?

(d) Are plastic hinges designed in columns, girders, or both? Explain your answer.

(e) Ties and special reinforcement are both used in concrete columns. Which is better? Why?

(f) How would you rate the effectiveness of stirrups in deep concrete beams? Explain your answer.

(g) What is the name given to the structural system which does not have a complete vertical load-carrying space frame? What K value is used for such a system?

(h) What percentage of the seismic load should be carried by a ductile moment-resisting space frame in a building taller than 160'?

(i) What percentage of the live load for warehouses and storage areas is considered for the W term in the formula $V = ZIKSCW$?

(j) What is the definition of rigidity?

Question 10

(a) What is the basic equation for "base shear?" Explain the meanings of the terms.

(b) Which of the terms in the equation for base shear can be equal to one? When are they be equal to one?

(c) What is the height limit above which a ductile moment-resisting space frame must be used?

(d) What are the various earthquake zones in California? What is the significance of these zones?

(e) What are the maximum accelerations and magnitudes for these zones?

(f) Describe the K values for a moment-resisting space frame.

(g) What K value would you use for a set of large grandstand bleachers at a football field?

(h) What are the forces in each column shown below? The top plate is rigid.

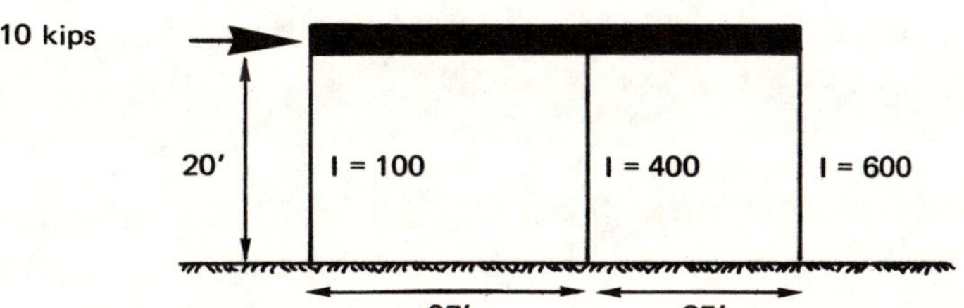

Question 11

(a)   A constant K is used in the formula for calculating base shear. Discuss the various uses, values, and applications of this constant.

(b)   What value of K would you use for a structure other than a building? List at least two situations.

(c)   The terms "ductility" and "flexibility" are often used in connection with seismic requirements. Are the two terms different? Give an example of how each is used in design.

(d)   In construction of concrete columns, it is common practice to use vertical bars with horizontal ties or spiral reinforcement. Which is more desireable, ties or spiral? Why?

(e)   How are building drift and the $P\Delta$ effect related? What are acceptable limits on drift?

NOTES

Bibliography

Ambrose, James, and Vergun, Dimitry, *Simplified Building Design for Wind and Earthquake Forces*, John Wiley & Sons, Inc., New York, NY, 1980

American Concrete Institute, "Building Code Requirements for Reinforced Concrete" (ACI 318-77), Detroit, MI, 1977

Amrhein, J.E., *Reinforced Masonry Engineering Handbook*, 2nd edition, Masonry Institute of America, Los Angeles, CA, 1973

Architectural Institute of Japan, (edited by) *Design Essentials in Earthquake Resistant Buildings*, Elsevier Publishing Company, New York, NY, 1970

Dowrick, D.J. *Earthquake Resistant Design*, John Wiley & Sons, Inc., New York, NY, 1977

Hamilton, Warren, "Plate Tectonics and Man," USGS Annual Report, Fiscal Year 1966, reprint available from the U.S. Government Printing Office, number 1978-261-227/27

Iacopi, Robert, *Earthquake Country*, Lane Book Company, Menlo Park, CA, 1964

International Conference of Building Officials, *Uniform Building Code*, Whittier, CA, 1979

Merritt, Frederic S., *Structural Steel Designers' Handbook*, McGraw-Hill Book Company, New York, NY, 1972

Seto, William W., *Mechanical Vibrations*, McGraw-Hill Book Company, New York, NY, 1964

Structural Engineers Association of California, "Recommended Lateral Force Requirements and Commentary," 217 Second Street, San Francisco, CA, 1980

Teal, Edward J., "Seismic Design Practice for Steel Buildings," published in *Engineering Journal*, American Institute of Steel Construction, Volume 12, No. 4, Fourth Quarter 1975

United States Geological Survey, "Earthquakes," publication 1979-311-348/11

United States Geological Survey, "The Severity of an Earthquake," publication 024-001-03204-0 (1979-0-281-363 (33))

United States Geological Survey, "The San Andreas Fault," publication USGS: Inf-66-3(R.10) stock #024-001-03015-2, (1977-240-966-11)

United States Geological Survey, "Active Faults in California," U.S. Government Printing Office #1977-0-240-96 6/48

Wiegel, Robert L., editor, *Earthquake Engineering*, Prentice-Hall, Englewood Cliffs, NJ, 1970

APPENDIX A: 1980 SEAOC CODE*

"Section 1 - General Requirements for the Design and Construction of Earthquake Resistive Structures"

(A) General

The proper application of these lateral force requirements, both in design and construction, are intended to provide minimum standards toward making buildings and other structures earthquake resistive. The provisions of this Section apply to the structure as a unit and also to all parts thereof, including the structural frame or walls, floor and roof systems, and other elements.

Every structure shall be designed and constructed to resist stresses produced by lateral forces as provided in this Section. Stresses shall be calculated as the effect of a force applied horizontally at each floor or roof level above the base. The force shall be assumed to come from any horizontal direction.

Where prescribed wind loads produce higher stresses, such loads shall be used in lieu of the loads resulting from earthquake forces.

(B) Definitions

BASE is the level at which the earthquake motions are considered to be imparted to the structure or the level at which the structure as a dynamic vibrator is supported.

BOX SYSTEM is a structural system without a complete vertical load carrying space frame. In this system, the required lateral forces are resisted by shear walls or braced frames as hereinafter defined.

BRACED FRAME is a truss system or its equivalent which is provided to resist lateral forces and in which the members are subjected primarily to axial stresses.

DUCTILE MOMENT RESISTING SPACE FRAME is a moment resisting space frame complying with the requirements given in Sections 2 and 4.

ESSENTIAL FACILITIES are those structures which must be functional for emergency post-earthquake operations.

LATERAL FORCE RESISTING SYSTEM is that part of the structural system assigned to resist the lateral forces prescribed in Section 1(D).

MOMENT RESISTING SPACE FRAME is a vertical load carrying space frame in which the members and joints are capable of resisting forces primarily by flexure.

SHEAR WALL is a wall designed to resist lateral forces parallel to the plane of the wall.

SPACE FRAME is a three-dimensional structural system, without bearing walls, composed of interconnected members laterally supported so as to function as a complete self-contained unit with or without the aid of horizontal diaphragms or floor bracing systems.

VERTICAL LOAD CARRYING SPACE FRAME is a space frame designed to carry all vertical loads.

(C) Symbols and Notations

The following symbols and notations apply to the provisions of this section.

C	numerical coefficient as specified in section 1(D)
C_p	numerical coefficient as specified in section 1(G) and as set forth in table 1-B
D	the dimension of the building in feet, in a direction parallel to the applied forces
δ_i	deflection at level i relative to the base, due to applied lateral forces, Σf_i, for use in formula (1-3)
f_i	distributed portion of a total lateral force at level i for use in formula (1-3)
F_i, F_n, F_x	lateral force applied to level i, n, or x respectively
F_p	lateral forces on a part of the structure and in the direction under consideration.
F_t	that portion of V considered concentrated at the top of the structure in addition to F_n.
g	acceleration due to gravity
h_i, h_n, h_x	height in feet above the base to level i, n, or x respectively
I	occupancy importance coefficient
K	numerical coefficient as set forth in table 1-A
Level i	Level of the structure referred to by the subscript i. i = 1 designates the first level above the base.
Level n	that level which is uppermost in the main portion of the structure
Level x	that level which is under design consideration. x = 1 designates the first level above the base.
N	the total number of stories above the base to level n
S	numerical coefficient for site-structure resonance
T	fundamental elastic period of vibration of the structure in seconds in the direction under consideration
T_s	characteristic site period
V	the total lateral force or shear at the base
W	the total dead load and applicable portions of other loads
w_i, w_x	that portion of W which is located at or is assigned to level i or x respectively
w_{px}	the weight of the diaphragm and the elements tributary thereto at level x, including 25% of the floor live load in storage and warehouse occupancies
W_p	the weight of a portion of a structure
Z	numerical coefficient related to the seismicity of a region

---

(D)   Minimum Earthquake Forces for Structures

Except as provided in Section 1(G) and 1(I), every structure shall be designed and constructed to resist minimum total lateral seismic forces assumed to act non-concurrently in the direction of each of the main axes of the structure in accordance with the formula

$$V = ZIKCSW \qquad (1\text{-}1)$$

The value of Z equals 1.0 for areas of highest seismicity.

The value of I equals 1.5 for essential facilities. For all others, I shall not be less than 1.0.

The value of K shall be not less than that set forth in Table 1-A.

The values of C and S are as indicated hereafter except that the product of CS need not exceed 0.14.

W is the total dead load and applicable portions of other loads such as: partitions, permanent equipment, snow, and, in storage and warehouse occupancies, a minimum of 25% of the floor live load.

The value of C shall be determined in accordance with the formula

$$C = \frac{1}{15\sqrt{T}} \qquad (1\text{-}2)$$

The value of C need not exceed 0.12.

The period T shall be established using the structural properties and deformational characteristics of resisting elements in a properly substantiated analysis such as the formula

$$T = 2\pi \sqrt{\left(\sum_{i=1}^{n} w_i\, \delta_i^2 \right) \div \left(g \sum_{i=1}^{n-1} f_i \delta_i \right)}$$

$$(1\text{-}3)$$

where the values of f_i represent any lateral force distributed approximately in accordance with the principles of formulas (1-5), (1-6), and (1-7) or any other rational distribution. The elastic deflections, δ_i, shall be calculated using the applied lateral forces, f_i.

In the absence of a period determination as indicated above, the value of T for buildings may be determined by the formula

$$T = \frac{0.05 h_n}{\sqrt{D}} \qquad (1\text{-}3A)$$

or, for buildings in which the lateral force resisting system consists of moment resisting space frames capable of resisting 100% of the required lateral forces and such system is not enclosed by or adjoined by more rigid elements tending to prevent the frame from resisting lateral forces, T may be determined by the formula

$$T = 0.10N \qquad (1\text{-}3B)$$

The value of S shall be determined by the following formulas but shall not be less than 1.0:

For $T/T_s = 1.0$ or less,

$$S = 1.0 + \frac{T}{T_s} - 0.5 \left[\frac{T}{T_s} \right]^2 \qquad (1\text{-}4)$$

For T/T_s greater than 1.0,

$$S = 1.2 + 0.6 \frac{T}{T_s} - 0.3 \left[\frac{T}{T_s} \right]^2 \qquad (1\text{-}4A)$$

T in Formulas (1-4) and (1-4A) shall be established by a properly substantiated analysis but T shall not be taken as less than 0.3 sec.

The range of values of T_s may be established from properly substantiated geotechnical data, except that T_s shall not be taken as less than 0.5 sec. nor more than 2.5 sec. T_s shall be that value within the range of site periods, as determined above, that is nearest to T.

When T_s is not properly established, the value of S shall be 1.5.

> *EXCEPTION:* Where T has been established by a properly substantiated analysis and exceeds 2.5 sec., the value of S may be determined by assuming a value of 2.5 sec. for T_s.

(E)   Distribution of Lateral Forces

1. Regular Structures or Framing Systems. The total lateral force V shall be distributed over the height of the structure in accordance with the following formulas:

$$V = F_t + \sum_{i=1}^{n} F_i \qquad (1\text{-}5)$$

The concentrated force at the top, F_t, shall be determined by the formula

$$F_t = 0.07\, TV \qquad (1\text{-}6)$$

F_t need not exceed $0.25V$ and may be considered as zero where T is 0.7 sec. or less. The remaining portion of the total base shear V shall be distributed over the height of the structure including level n according to the formula

$$F_x = \frac{(V - F_t)w_x h_x}{\sum_{i=1}^{n} w_i h_i} \qquad (1\text{-}7)$$

At each level designated as x, the force F_x shall be applied over the area of the building in accordance with the mass distribution on that level.

2. Setbacks. Buildings having setbacks wherein the plan dimension of the tower in each direction is at least 75% of the corresponding plan dimension of the lower part may be considered as uniform buildings without setbacks, providing other irregularities as defined in this Section do not exist.

3. Irregular Structures or Framing Systems. The distribution of the lateral forces in structures which have highly irregular shapes, large differences in lateral resistance or stiffness between adjacent stories or other unusual structural features shall be determined considering the dynamic characteristics of the structure.

4. Distribution of Horizontal Shear. Total shear in any horizontal plane shall be distributed to the various elements of the lateral force resisting system in proportion to their rigidities, considering the rigidity of the horizontal bracing system or diaphragm. Rigid elements that are assumed not to be part of the lateral force-resisting system may be incorporated into buildings provided that their effect on the action of the system is considered and provided for in the design.

5. Horizontal Torsional Moments. Provisions shall be made for the increase in shear resulting from the horizontal torsion due to an eccentricity between the center of mass and the center of rigidity. Negative torsional shears shall be neglected. Where the vertical resisting elements depend on diaphragm action for shear distribution at any level, the shear resisting elements shall be capable of resisting a torsional moment assumed to be equivalent to the story shear acting with an eccentricity of not less than 5% of the maximum building dimension at that level.

(F) Overturning

Every structure shall be designed to resist the overturning effects caused by the wind forces and related requirements, or the earthquake forces specified in this Section, whichever governs.

At any level, the incremental changes of the design overturning moment, in the story under consideration, shall be distributed to the various resisting elements in the same proportion as the distribution of the shears in the resisting system. Where other vertical members are provided which are capable of partially resisting the overturning moments, a redistribution may be made to these members if framing members of sufficient strength and stiffness to transmit the required loads are provided.

Where a vertical resisting element is discontinuous, the overturning moment carried by the lowest story of that element shall be carried down as loads to the foundation.

(G) Lateral Force on Elements of Structures

Parts or portions of structures and their anchorage to the main structural system shall be designed for lateral forces in accordance with the formula

$$F_p = Z\, I\, C_p\, W_p \qquad (1\text{-}8)$$

The values of C_p are set forth in Table 1-B. The values of the I coefficient shall be the value used for the building.

EXCEPTIONS:
1. The value of I for panel connectors shall be as given in section 1(J) 3.d.
2. The value of I for elements of life safety systems shall be 1.5.

The distribution of these forces shall be according to the gravity loads pertaining thereto.
For applicable forces on diaphragms and connections for exterior panels refer to sections 1(J) 2.d and 1(J) 3.d, respectively.

(H) Drift Provisions

1. Drift. Lateral deflections or drift of a story relative to its adjacent stories shall not exceed 0.005 times the story height unless it can be demonstrated that greater drift can be tolerated. The displacement calculated from the application of the required lateral forces shall be multiplied by $(1.0/K)$ to obtain the drift. The ratio $(1.0/K)$ shall not be less than 1.0.

2. Building Separations. All portions of structures shall be designed and constructed to act as an integral unit in resisting horizontal forces unless separated structurally by a distance sufficient to avoid contact under deflection from seismic action or wind forces.

(I) Alternate Determination and Distribution of Seismic Forces

Nothing in these Recommendations shall be deemed to prohibit the submission of properly substantiated technical data for establishing the lateral design forces and distribution by dynamic analyses. In such analyses the dynamic characteristics of the structure must be considered.

(J) Structural Systems

1. Ductility Requirements

a. Force Factor. All buildings designed with a horizontal force factor $K = 0.67$ or 0.80 shall have ductile moment resisting space frames.

b. Tall Buildings. In zones of high seismicity, buildings more than one hundred and sixty feet (160') in height shall have ductile moment resisting space frames capable of resisting not less than 25% of the required seismic forces for the structure as a whole.

c. Concrete Fràmes. In zones of high seismicity, all concrete space frames required by design to be part of the lateral force resisting system and all concrete frames located in the perimeter line of vertical support shall be ductile moment resisting space frames.

EXCEPTION: Frames in the perimeter line of vertical support of buildings designed with shear walls taking 100% of the design lateral forces need only conform with Section 1(J)1 d.

d. Deformation Compatibility. All framing elements not required by design to be part of the lateral force resisting system shall be investigated and shown to be adequate for vertical load carrying capacity and induced moments due to $(3.0/K)$ times the distortions resulting from the required lateral forces. The rigidity of other elements shall be considered in accordance with Section 1(E)4.

e. Adjoining Rigid Elements. Moment resisting space frames and ductile moment resisting space frames may be enclosed by or adjoined by more rigid elements which would tend to prevent the space frame from resisting lateral forces where it can be shown that the action or failure of the more rigid elements will not impair the vertical and lateral load resisting ability of the space frame.

f. Frame Ductility. The necessary ductility for a ductile moment resisting space frame shall be provided by a frame of structural steel conforming to Section 4, or by a reinforced concrete frame complying with Section 2 of these Recommendations.

g. Braced Frames. All members in braced frames shall be designed for 1.25 times the force determined in accordance with Section 1(D). Connections shall be designed to develop the full capacity of the members or shall be based on the above forces without the one-third increase usually permitted for stresses resulting from earthquake forces. Members of braced frames shall be composed of ASTM A36, A441, A500 (Grades B and C), A501, A572 (Grades 42, 45, 50, 55) or A588 structural steel; or reinforced concrete bracing members conforming with the requirements of section 3(B) of the Recommendations.

h. Shear Walls. Reinforced concrete shear walls for all structures shall conform to the requirements of Section 3 of these Recommendations. For the calculation of shear stress only, all masonry shear walls shall be designed to resist 1.5 times the force determined in accordance with Section 1(D).

i. Framing Below Base. In buildings where $K = 0.67$ or 0.80, the special ductility requirements of Sections 2, 3, and 4 of these Recommendations, as appropriate, shall apply to all structural elements below the base which are required to transmit to the foundation the forces resulting from lateral loads.

2. Design Requirements.

a. Minor Alterations. Minor structural alterations may be made in existing buildings and other structures, but the resistance to lateral forces shall be not less than that before such alterations were made unless the building as altered meets the requirements of these Recommendations.

b. Reinforced Masonry or Concrete. In zones of high seismicity, all elements within the structure which are of masonry or concrete shall be reinforced so as to qualify as reinforced masonry or concrete.

c. Combined Vertical and Horizontal Forces. In computing the effect of seismic forces in combination with vertical loads, gravity load stresses induced in members by dead load plus design live load, except roof live load, shall be considered. Consideration should also be given to minimum gravity loads acting in combination with lateral forces.

d. Diaphragms. Floor and roof diaphragms shall be designed to resist the seismic forces determined in accordance with the following formula:

$$F_{px} = \left[\frac{F_t + \sum_{i=x}^{n} F_i}{\sum_{i=x}^{n} w_i} \right] w_{px} \qquad (1\text{-}9)$$

The force F_{px} determined from Formula 1-9 need not exceed $0.30\, Z\, I\, w_{px}$.

The diaphragm may be required to transfer seismic forces from the vertical resisting elements above the diaphragm to other vertical resisting elements below the diaphragm. These forces shall be added to those determined from Formula 1-9. These additional forces may be due to offsets or changes in stiffness in the vertical elements.

However, in no case shall the seismic force on the diaphragm be less than determined by the following formula:

$$F_{px} = 0.14\, Z\, I\, w_{px} \qquad (1\text{-}9A)$$

Diaphragms supporting concrete or masonry walls shall have continuous ties between diaphragm chords to distribute the anchorage forces specified in Section 1(J)3a into the diaphragm. Added chords may be used to form sub-diaphragms to transmit the anchorage forces to the main cross ties. Diaphragm deformations shall be considered in the design of the supported walls. (See Section 1(J)3b for special anchorage requirements of wood diaphragms.)

3. Special Requirements.

a. Anchorage of Concrete or Masonry Walls. Concrete or masonry walls shall be anchored to all floors and roofs which provide lateral support for the wall. The anchorage shall provide a positive direct connection between the walls and floor or roof construction capable of resisting the horizontal forces specified in these Recommendations or a minimum force of 200 pounds per lineal foot of wall, whichever is greater. Walls shall be designed to resist bending between anchors where the anchor spacing exceeds four feet. In masonry walls of hollow units or cavity walls, anchors shall be embedded in a reinforced grouted structural element of the wall. (See Section 1(J)2d for the requirements for developing anchorage forces in diaphragms. See Section 1(J)3b for special anchorage requirements for wood diaphragms.)

b. Wood Diaphragms Used to Support Concrete or Masonry Walls. Where wood diaphragms are used to laterally support concrete or masonry walls the anchorage shall conform to Section 1(J)3a. In zones of high seismicity, anchorage shall not be accomplished by use of toe nails, or nails subjected to withdrawal; nor shall wood ledgers be used in cross grain bending. The continuous ties required by Section 1(J)2d shall be in addition to the diaphragm sheathing; the diaphragm sheathing shall not be used to splice these ties.

c. Pile Caps and Caissons. Individual pile caps and caissons of every building or structure shall be interconnected by ties, each of which can carry by tension and compression a minimum horizontal force equal to 10 percent of the larger column loading, unless it can be demonstrated that equivalent restraint can be provided by other approved methods.

d. Exterior Elements. Precast or prefabricated nonbearing, nonshear wall panels or similar elements which are attached to or enclose the exterior, shall be designed to resist the forces per Formula (1-8) and shall accomodate movements of the structure resulting from lateral forces or temperature changes. Concrete panels or other similar elements shall be supported by means of cast-in-place concrete or by mechanical connections and fasteners in accordance with the following provisions:

Connections and panel joints shall allow for a relative movement between stories of not less than two times story drift caused by wind or (3.0/K) times the calculated elastic story displacement caused by required seismic forces, or ½ inch, whichever is greater.

Connections to permit movement in the plane of the panel for story drift shall be properly designed sliding connections using slotted or oversize holes or may be connections which permit movement by bending of steel or other connections providing equivalent sliding and ductility capacity.

Bodies of connections shall have sufficient ductility and rotation capacity so as to preclude fracture of the concrete or brittle failures at or near welds.

The body of the connection shall be designed for 1-1/3 times the force determined by Formula (1-8).

Elements connecting the bodies to the panel or the structure such as bolts, inserts, welds, dowels, etc., shall be designed for 4 times the forces determined by Formula (1-8).

Elements of connections embedded in concrete shall be attached to, or hooked around reinforcing steel, or otherwise terminated so as to effectively transfer forces to the reinforcing steel.

The value of the coefficient I shall be 1.0 for the entire connection.

Table 1-A. Horizontal Force Factor K for Buildings or Other Structures

Type or Arrangement of Resisting Elements	Value of K
All building framing systems except as hereinafter classified.	1.00
Building with a box system as defined in Section 1(B).	1.33

EXCEPTION: Buildings not more than three stories in height with stud-wall framing and using horizontal diaphragms and vertical shear panels for the lateral force system, may use K = 1.0.

Type or Arrangement of Resisting Elements	Value of K
Buildings with a dual bracing system consisting of a ductile moment resisting space frame and shear walls or braced frames designed in accordance with the following criteria: 1. The frame and shear walls or braced frames shall resist the total lateral force in accordance with their relative rigidities considering the interaction of the shear walls and frames. 2. The shear walls or braced frames acting independently of the ductile moment resisting space frame shall resist the total required lateral force. 3. The ductile moment resisting space frame shall have the capacity to resist not less than 25 percent of the required lateral force.	0.80
Buildings with a ductile moment resisting space frame designed in accordance with the following criteria: The ductile moment resisting space frame shall have the capacity to resist the total required lateral force.	0.67
Elevated tanks plus full contents, on four or more cross-braced legs and not supported by a building.[1,2]	2.5
Structures other than buildings and other than those set forth in Table 1-B.	2.0

[1] See Sect. 1(J)1g for additional detail requirements.

[2] The torsional requirements of Sect. 1(E)5 shall apply.

Table 1-B. Horizontal Force Factor C_p for Elements of Structures

Cantilevered Elements a. Parapets	Normal to flat surfaces	.8
b. Chimneys or stacks	Any direction	
All other walls, partitions, and similar elements — see also Sec. 1(J) 3.d	Any direction	.3
Exterior and interior ornamentations and appendages	Any direction	.8

When connected to, part of, or housed within a building:
   a. Penthouses, anchorage and supports for tanks, including contents, chimneys and stacks
   b. Storage racks plus contents
   c. Suspended ceilings (3) .
   d. All equipment or machinery
	Any direction	.3[2][4]

Connections for prefabricated structural elements other than walls, with force applied at center of gravity of assembly	Any direction	0.3[4]

(1) C_p for elements laterally self supported only at ground level may be 2/ 3 of the value shown.

(2)   For flexible and flexibly mounted equipment and machinery, the appropriate values of C_p shall be determined with consideration given to both the dynamic properties of the equipment and machinery and to the building or structure in which it is placed but shall not be less than the listed values. The design of the equipment and machinery and their anchorage is an integral part of the design and specification of such equipment and machinery.

For Essential Facilities and life safety systems, the design and detailing of equipment which must remain in place and be functional following a major earthquake shall consider the effect of drift.

(3)   Ceiling weight shall include all light fixtures and other equipment or partitions which are laterally supported by the ceilings. For purposes of determining the lateral force, a ceiling weight of not less than 4 pounds per square foot shall be used.

(4)   The force shall be resisted by positive anchorage and not by friction.

Rigidity of Fixed Piers
Using F = 100,000, t = 1.0, and E = 1,000,000

(h/d)	.00	.01	.02	.03	.04	.05	.06	.07	.08	.09
0.10	33.223	30.181	27.645	25.497	23.655	22.057	20.657	19.421	18.321	17.335
0.20	16.447	15.643	14.911	14.242	13.627	13.061	12.538	12.053	11.602	11.181
0.30	10.788	10.419	10.073	9.747	9.440	9.150	8.876	8.616	8.369	8.135
0.40	7.911	7.699	7.496	7.302	7.117	6.939	6.769	6.606	6.449	6.299
0.50	6.154	6.015	5.880	5.751	5.626	5.506	5.389	5.277	5.168	5.062
0.60	4.960	4.862	4.766	4.673	4.583	4.495	4.410	4.328	4.247	4.169
0.70	4.093	4.019	3.948	3.877	3.809	3.743	3.678	3.615	3.553	3.493
0.80	3.434	3.377	3.321	3.266	3.213	3.160	3.109	3.060	3.011	2.963
0.90	2.916	2.871	2.826	2.782	2.739	2.697	2.656	2.616	2.577	2.538
1.00	2.500	2.463	2.427	2.391	2.356	2.322	2.288	2.255	2.222	2.191
1.10	2.159	2.129	2.099	2.069	2.040	2.012	1.984	1.956	1.929	1.903
1.20	1.877	1.851	1.826	1.802	1.777	1.753	1.730	1.707	1.684	1.662
1.30	1.640	1.619	1.598	1.577	1.556	1.536	1.516	1.497	1.478	1.459
1.40	1.440	1.422	1.404	1.386	1.369	1.352	1.335	1.318	1.302	1.286
1.50	1.270	1.254	1.239	1.224	1.209	1.194	1.180	1.166	1.152	1.138
1.60	1.124	1.111	1.098	1.085	1.072	1.059	1.047	1.034	1.022	1.010
1.70	.999	.987	.976	.965	.954	.943	.932	.921	.911	.901
1.80	.890	.880	.870	.861	.851	.842	.832	.823	.814	.805
1.90	.796	.788	.779	.771	.762	.754	.746	.738	.730	.722
2.00	.714	.707	.699	.692	.685	.677	.670	.663	.656	.649
2.10	.643	.636	.629	.623	.617	.610	.604	.598	.592	.586
2.20	.580	.574	.568	.562	.557	.551	.546	.540	.535	.530
2.30	.525	.519	.514	.509	.504	.499	.495	.490	.485	.480
2.40	.476	.471	.467	.462	.458	.453	.449	.445	.441	.437
2.50	.432	.428	.424	.420	.417	.413	.409	.405	.401	.398
2.60	.394	.391	.387	.383	.380	.377	.373	.370	.367	.363
2.70	.360	.357	.354	.350	.347	.344	.341	.338	.335	.332
2.80	.330	.327	.324	.321	.318	.316	.313	.310	.307	.305
2.90	.302	.300	.297	.295	.292	.290	.287	.285	.283	.280
3.00	.278	.276	.273	.271	.269	.267	.264	.262	.260	.258
3.10	.256	.254	.252	.250	.248	.246	.244	.242	.240	.238
3.20	.236	.234	.232	.231	.229	.227	.225	.223	.222	.220
3.30	.218	.217	.215	.213	.212	.210	.208	.207	.205	.204
3.40	.202	.201	.199	.198	.196	.195	.193	.192	.190	.189
3.50	.187	.186	.185	.183	.182	.181	.179	.178	.177	.175
3.60	.174	.173	.172	.170	.169	.168	.167	.166	.164	.163
3.70	.162	.161	.160	.159	.157	.156	.155	.154	.153	.152
3.80	.151	.150	.149	.148	.147	.146	.145	.144	.143	.142
3.90	.141	.140	.139	.138	.137	.136	.135	.134	.133	.132
4.00	.132	.131	.130	.129	.128	.127	.126	.126	.125	.124
4.10	.123	.122	.122	.121	.120	.119	.118	.118	.117	.116
4.20	.115	.115	.114	.113	.112	.112	.111	.110	.110	.109
4.30	.108	.108	.107	.106	.106	.105	.104	.104	.103	.102
4.40	.102	.101	.100	.100	.099	.099	.098	.097	.097	.096
4.50	.096	.095	.094	.094	.093	.093	.092	.092	.091	.091
4.60	.090	.089	.089	.088	.088	.087	.087	.086	.086	.085
4.70	.085	.084	.084	.083	.083	.082	.082	.081	.081	.080
4.80	.080	.080	.079	.079	.078	.078	.077	.077	.076	.076
4.90	.076	.075	.075	.074	.074	.073	.073	.073	.072	.072
5.00	.071	.071	.071	.070	.070	.069	.069	.069	.068	.068
5.10	.068	.067	.067	.066	.066	.066	.065	.065	.065	.064
5.20	.064	.064	.063	.063	.063	.062	.062	.062	.061	.061
5.30	.061	.060	.060	.060	.059	.059	.059	.058	.058	.058
5.40	.058	.057	.057	.057	.056	.056	.056	.056	.055	.055
5.50	.055	.054	.054	.054	.054	.053	.053	.053	.052	.052
5.60	.052	.052	.051	.051	.051	.051	.050	.050	.050	.050
5.70	.049	.049	.049	.049	.048	.048	.048	.048	.048	.047
5.80	.047	.047	.047	.046	.046	.046	.046	.045	.045	.045
5.90	.045	.045	.044	.044	.044	.044	.044	.043	.043	.043
6.00	.043	.043	.042	.042	.042	.042	.042	.041	.041	.041
6.10	.041	.041	.040	.040	.040	.040	.040	.039	.039	.039
6.20	.039	.039	.039	.038	.038	.038	.038	.038	.038	.037
6.30	.037	.037	.037	.037	.037	.036	.036	.036	.036	.036
6.40	.036	.035	.035	.035	.035	.035	.035	.034	.034	.034
6.50	.034	.034	.034	.034	.033	.033	.033	.033	.033	.033
6.60	.033	.032	.032	.032	.032	.032	.032	.032	.031	.031
6.70	.031	.031	.031	.031	.031	.031	.030	.030	.030	.030
6.80	.030	.030	.030	.029	.029	.029	.029	.029	.029	.029
6.90	.029	.029	.028	.028	.028	.028	.028	.028	.028	.028
7.00	.027	.027	.027	.027	.027	.027	.027	.027	.027	.026
7.10	.026	.026	.026	.026	.026	.026	.026	.026	.026	.025
7.20	.025	.025	.025	.025	.025	.025	.025	.025	.025	.024
7.30	.024	.024	.024	.024	.024	.024	.024	.024	.024	.023
7.40	.023	.023	.023	.023	.023	.023	.023	.023	.023	.023
7.50	.023	.022	.022	.022	.022	.022	.022	.022	.022	.022
7.60	.022	.022	.021	.021	.021	.021	.021	.021	.021	.021
7.70	.021	.021	.021	.021	.021	.020	.020	.020	.020	.020
7.80	.020	.020	.020	.020	.020	.020	.020	.020	.019	.019
7.90	.019	.019	.019	.019	.019	.019	.019	.019	.019	.019
8.00	.019	.019	.019	.018	.018	.018	.018	.018	.018	.018
8.10	.018	.018	.018	.018	.018	.018	.018	.018	.017	.017
8.20	.017	.017	.017	.017	.017	.017	.017	.017	.017	.017

Rigidity of Cantilever Piers
Using F = 100,000, t = 1.0, and 1,000,000

(h/d	.00	.01	.02	.03	.04	.05	.06	.07	.08	.09
0.10	32.895	29.822	27.255	25.076	23.203	21.575	20.146	18.880	17.752	16.738
0.20	15.823	14.992	14.233	13.538	12.898	12.308	11.761	11.252	10.778	10.335
0.30	9.921	9.531	9.165	8.820	8.495	8.187	7.895	7.618	7.356	7.106
0.40	6.868	6.642	6.425	6.219	6.021	5.833	5.652	5.479	5.313	5.153
0.50	5.000	4.853	4.712	4.576	4.445	4.319	4.197	4.080	3.968	3.859
0.60	3.754	3.652	3.555	3.460	3.369	3.280	3.195	3.112	3.032	2.955
0.70	2.880	2.808	2.738	2.670	2.604	2.540	2.478	2.418	2.360	2.303
0.80	2.248	2.195	2.143	2.093	2.045	1.997	1.952	1.907	1.864	1.822
0.90	1.781	1.741	1.702	1.665	1.628	1.593	1.558	1.524	1.492	1.460
1.00	1.429	1.398	1.369	1.340	1.312	1.285	1.259	1.233	1.208	1.183
1.10	1.160	1.136	1.114	1.092	1.070	1.049	1.028	1.008	.989	.970
1.20	.951	.933	.916	.898	.881	.865	.849	.833	.818	.803
1.30	.788	.774	.760	.746	.733	.720	.707	.695	.683	.671
1.40	.659	.648	.636	.626	.615	.604	.594	.584	.575	.565
1.50	.556	.546	.537	.529	.520	.512	.503	.495	.487	.480
1.60	.472	.465	.457	.450	.443	.436	.430	.423	.417	.410
1.70	.404	.398	.392	.386	.380	.375	.369	.364	.358	.353
1.80	.348	.343	.338	.333	.329	.324	.319	.315	.310	.306
1.90	.302	.298	.294	.290	.286	.282	.278	.274	.270	.267
2.00	.263	.260	.256	.253	.250	.246	.243	.240	.237	.234
2.10	.231	.228	.225	.222	.219	.216	.214	.211	.208	.206
2.20	.203	.201	.198	.196	.194	.191	.189	.187	.184	.182
2.30	.180	.178	.176	.174	.172	.170	.168	.166	.164	.162
2.40	.160	.158	.156	.155	.153	.151	.149	.148	.146	.145
2.50	.143	.141	.140	.138	.137	.135	.134	.132	.131	.129
2.60	.128	.127	.125	.124	.123	.121	.120	.119	.118	.116
2.70	.115	.114	.113	.112	.111	.109	.108	.107	.106	.105
2.80	.104	.103	.102	.101	.100	.099	.098	.097	.096	.095
2.90	.094	.093	.092	.091	.091	.090	.089	.088	.087	.086
3.00	.086	.085	.084	.083	.082	.082	.081	.080	.079	.079
3.10	.078	.077	.076	.076	.075	.074	.074	.073	.072	.072
3.20	.071	.071	.070	.069	.069	.068	.067	.067	.066	.066
3.30	.065	.065	.064	.063	.063	.062	.062	.061	.061	.060
3.40	.060	.059	.059	.058	.058	.057	.057	.056	.056	.055
3.50	.055	.055	.054	.054	.053	.053	.052	.052	.052	.051
3.60	.051	.050	.050	.050	.049	.049	.048	.048	.048	.047
3.70	.047	.046	.046	.046	.045	.045	.045	.044	.044	.044
3.80	.043	.043	.043	.042	.042	.042	.041	.041	.041	.040
3.90	.040	.040	.040	.039	.039	.039	.038	.038	.038	.038
4.00	.037	.037	.037	.037	.036	.036	.036	.035	.035	.035
4.10	.035	.034	.034	.034	.034	.034	.033	.033	.033	.033
4.20	.032	.032	.032	.032	.031	.031	.031	.031	.031	.030
4.30	.030	.030	.030	.030	.029	.029	.029	.029	.029	.028
4.40	.028	.028	.028	.028	.028	.027	.027	.027	.027	.027
4.50	.026	.026	.026	.026	.026	.026	.025	.025	.025	.025
4.60	.025	.025	.024	.024	.024	.024	.024	.024	.024	.023
4.70	.023	.023	.023	.023	.023	.023	.022	.022	.022	.022
4.80	.022	.022	.022	.021	.021	.021	.021	.021	.021	.021
4.90	.021	.020	.020	.020	.020	.020	.020	.020	.020	.020
5.00	.019	.019	.019	.019	.019	.019	.019	.019	.019	.018
5.10	.018	.018	.018	.018	.018	.018	.018	.018	.017	.017
5.20	.017	.017	.017	.017	.017	.017	.017	.017	.017	.016
5.30	.016	.016	.016	.016	.016	.016	.016	.016	.016	.016
5.40	.015	.015	.015	.015	.015	.015	.015	.015	.015	.015
5.50	.015	.015	.015	.014	.014	.014	.014	.014	.014	.014
5.60	.014	.014	.014	.014	.014	.014	.013	.013	.013	.013
5.70	.013	.013	.013	.013	.013	.013	.013	.013	.013	.013
5.80	.013	.012	.012	.012	.012	.012	.012	.012	.012	.012
5.90	.012	.012	.012	.012	.012	.012	.012	.012	.012	.012
6.00	.011	.011	.011	.011	.011	.011	.011	.011	.011	.011
6.10	.011	.011	.011	.011	.011	.011	.010	.010	.010	.010
6.20	.010	.010	.010	.010	.010	.010	.010	.010	.010	.010
6.30	.010	.010	.010	.010	.010	.010	.010	.009	.009	.009
6.40	.009	.009	.009	.009	.009	.009	.009	.009	.009	.009
6.50	.009	.009	.009	.009	.009	.009	.009	.009	.009	.009
6.60	.009	.009	.008	.008	.008	.008	.008	.008	.008	.008
6.70	.008	.008	.008	.008	.008	.008	.008	.008	.008	.008
6.80	.008	.008	.008	.008	.008	.008	.008	.008	.008	.008
6.90	.007	.007	.007	.007	.007	.007	.007	.007	.007	.007
7.00	.007	.007	.007	.007	.007	.007	.007	.007	.007	.007
7.10	.007	.007	.007	.007	.007	.007	.007	.007	.007	.007
7.20	.007	.007	.007	.007	.006	.006	.006	.006	.006	.006
7.30	.006	.006	.006	.006	.006	.006	.006	.006	.006	.006
7.40	.006	.006	.006	.006	.006	.006	.006	.006	.006	.006
7.50	.006	.006	.006	.006	.006	.006	.006	.006	.006	.006
7.60	.006	.006	.006	.006	.006	.006	.005	.005	.005	.005
7.70	.005	.005	.005	.005	.005	.005	.005	.005	.005	.005
7.80	.005	.005	.005	.005	.005	.005	.005	.005	.005	.005
7.90	.005	.005	.005	.005	.005	.005	.005	.005	.005	.005
8.00	.005	.005	.005	.005	.005	.005	.005	.005	.005	.005
8.10	.005	.005	.005	.005	.005	.005	.005	.005	.005	.005
8.20	.004	.004	.004	.004	.004	.004	.004	.004	.004	.004

THE PRACTICE OF STRUCTURAL ENGINEERING

Reprinted from "The San Fernando Earthquake of February 9, 1971 and Public Policy,'" (pages 47-49), by the Special Subcommittee of the Joint Committee on Seismic Safety - California Legislature, July 1972.

Under existing laws of the State of California, buildings may legally be designed for earthquake resistance by the four following classifications of design professionals:

1. Under Chapter 3, Division 3 of the Business and Professions Code, a registered architect may perform and take responsibility for the structural design work for buildings of any type or size without restriction. Most architects do receive a nominal amount of academic training in structural engineering and all architects are required to pass a rather elementary examination in structural engineering as part of their licensing examination. It is true that most knowledgeable architects do use structural engineers either on their staff or as consultants to perform the structural design work required for their building projects. Nevertheless, legally, any licensed architect is permitted to perform and assume responsibility for structural designs of any complexity.

> 5500.1 *Practice of architecture*
>
> A person engages in the practice of architecture within the meaning and intent of this chapter who holds himself out as able to perform or who does perform any service which requires or would require the application of the science, art, or profession of planning sites or of planning or designing buildings or architectural structures and their related facilities. Such services may include consultation, investigation, evaluation, planning, design, the preparation of instruments of service such as drawings and specifications, and supervision of construction insofar as customarily performed by architects. (Added Stats. 1963, c. 2133, p. 4433, § 3.)

2. Under Chapter 7 of Division 3 of the Business and Professions Code, a registered civil engineer may assume responsibility for the structural design work for any building regardless of its magnitude or complexity. Some governmental jurisdictions by their own codes or regulations do require that only those civil engineers with a structural engineer's title can be responsible for the structural design and supervision of construction of buildings above a certain size. The State Education Code under Section 15459 requires that structural designs for public school buildings be prepared by either a certified architect or structural engineer. The Section reads as follows:

> 15459. All plans, specifications, and estimates shall be prepared by a certified architect holding a valid license under Chapter 3 of Division 3 of the Business and Professions Code or by a structural engineer holding a valid certificate to use the title structural engineer under Chapter 7 of Division 3 of the Business and Professions Code,

and the supervision of the work of construction shall be under the responsible charge of such an architect or structural engineer, except that where plans, specifications, and estimates for alterations or repairs do not involve architectural or structural changes said plans, specifications, and estimates may be prepared and work of construction may be supervised by a professional engineer duly qualified to perform such services and holding a valid certificate under Chapter 7 of Division 3 of the Business and Professions Code for performance of services in that branch of engineering in which said plans, specifications, and estimates and work of construction are applicable.

Most civil engineers in their academic training do take a certain number of basic courses which are fundamental to structural engineering. Also, all registered civil engineers are required to pass a modest examination covering basic structural engineering principles and some practical building design as part of a complete examination which covers a very wide range of diversified civil engineering fields described in the law as follows:

> 6731. Civil engineering embraces the following studies or activities in connection with fixed works for irrigation, drainage, waterpower, water supply, flood control, inland waterways, harbors, municipal improvements, railroads, highways, tunnels, airports and airways, purification of water, sewerage, refuse disposal, foundations, framed and homogeneous structures, buildings, or bridges:
>
> (a) The economics of, the use and design of, materials of construction and the determination of their physical qualities.
> (b) The supervision of the construction of engineering structures.
> (c) The investigation of the laws, phenomena and forces of nature.
> (d) Appraisals or valuations.
> (e) The preparation and/or submission of designs, plans and specifications and engineering reports.
>
> Civil engineering also includes city and regional planning insofar as any of the above features are concerned therein, and land surveying as defined in Chapter 15 (commencing at Section 8700) of Division 3.

Legally, a registered civil engineer is permitted to assume responsibility for structural design work of any complexity (except for the exceptions noted) without any further experience or qualifying examination in earthquake design principles. As an example, a young civil engineer two years out of college could conceivably take responsibility for the design of a hospital or high-rise building of any complexity.

3. Under Chapter 7 of Division 3 of the Business and Professions Code, a registered civil engineer may become authorized to use the title of Structural Engi-

neer after three years of work under the supervision of a registered structural engineer and after a comprehensive 16 hour examination mainly concerned with earthquake engineering design principles and practical construction details. A registered structural engineer is, of course, permitted to assume responsibility for structural engineering design work for any building and is required by law to do this work on all public school buildings unless the responsibility for the structural design is assumed by an architect. The registration of structural engineers by the State of California was authorized specially as a means of obtaining the best professional talent in designing earthquake resistant buildings and hence reducing seismic risk.

> 6736. No person shall use the title, "structural engineer," unless he is a registered civil engineer in this State and, furthermore, unless he has been found qualified as a structural engineer according to the rules and regulations established therefor by the board.
>
> (4) The examination for authority to use the title of structural engineer shall be held annually in August or September. It shall consist of a two-day examination, requiring a demonstration of special knowledge and experience in structural engineering. The field to be covered may be broad and may include the design and detailing of all types of structures, subject to both gravity and horizontal loads.

4. Under Chapter 3, Division 3 of the Business and Professions Code, a registered building designer may provide the same engineering services as an architect, structural engineer, or civil engineer by associating with any of these professionals, who shall assume all legal responsibilities involved.

> 5500.2 *Building designer*
> As used in this chapter, building designer means a person who has been registered to engage in the practice of building design in this State under the authority of this chapter, and who is a potential candidate for certification as an architect.
>
> 5500.3 *Practice of building design*
> A person engages in the practice of building design, within the meaning and intent of this chapter, who holds himself out as able to perform or who does perform any service which requires or would require the application of the skills necessary for designing buildings and their related sites and facilities, and includes the preparation of customary instruments of service such as drawings and specifications and supervision of construction insofar as is customarily performed by building designers.
>
> A person who is registered as a building designer shall associate with a registered civil engineer, structural engineer, or licensed architect with respect to the design of any building which is not exempted from Chapter 7 (commencing with Section 6700) by Section 6737.1; and such association shall prepare or offer to prepare drawings, specifications, estimates, or instruments of service for the construction of buildings where such documents are required by law to be prepared by a registered civil engineer, structural engineer or licensed architect. The

registered civil engineer, structural engineer, or licensed architect shall sign all drawings, specifications, or instruments of service which shall be evidence of his responsibility for the work. The registered civil engineer, structural engineer, or licensed architect shall be responsible for the supervision of construction.

> A building designer may not prepare or offer to prepare drawings, specifications, estimates or instruments of service with respect to the design of any building which is not exempted from Chapter 7 (commencing with Section 6700 by Section 6737.1 unless the building designer is associated with a registered civil engineer, structural engineer, or licensed architect prior to offering such services.

From the above, it is apparent that there are serious weaknesses in State laws regulating the practice of structural engineering which allow unqualified professionals to perform structural engineering work of any complexity.

It is strongly urged that serious consideration be given to new or amended legislation which would require that registered structural engineers perform earthquake-resistant design for all except minor structures within the state. It is realized that this is a very far-reaching recommendation. However, the problem should be considered if the state is to make a serious effort to take all practical steps to improve seismic safety.

Answers to SEISMIC DESIGN questions

NOTE: Many of the questions are answered by referrals to a page in the SEISMIC DESIGN text. This is done whenever the text addresses the question directly. The answer is given in greater detail if the text does not cover the question or if additional detail is necessary.

QUESTION 1

(a) In general, steel at any point in a structure will have failed if it is stressed beyond the steel's yield point. This can be expected to occur:
(1) at the ends of beams and girders with the formation of plastic hinges
(2) in columns due to buckling
(3) in columns due to bending (the P-Δ effect)
(4) in flanges and webs due to excess stress reversal (fatigue loading)
(5) in girder-column connections that cannot sustain the full plastic moment of a connecting member

Other non-structural failures can be expected to occur since steel structures are lightweight and responsive to earthquake oscillations. Examples of non-strucutral failures are: damaged pipelines, broken glass, cracked filler walls, and stairwell shifts.

(b) In general, a reinforced concrete frame will have failed if the concrete spalls (crushes) before plastic yielding of the steel reinforcing occurs, or if the steel reinforcing is stressed plastically. Failure can be expected to occur:
(1) in adequately-designed columns, at the ends when there is insufficient diagonal resistance (column sidesway)
(2) in poorly-designed columns with insufficient confinement
(3) in shear walls due to inadequate vertical reinforcing
(4) at construction joints due to poor bonding (typical)
(5) inadequate shear reinforcing in beams (as between shear walls)
(6) in columns due to excessive overturning moment

(c) See pages 53 and 54. The strength method should be used to allow concrete members to achieve plastic performance.
(d) See pages 62 to 64. The most important consideration is confinement of concrete. Confined reinforced concrete will fail in the steel before the concrete crushes. Adequate bonding between the steel and the concrete must be maintained. Similarly, all splices should be welded to maintain development lengths.
(e) Stiffness is the amount of force required to cause a unit of displacement. (See "k" in example 2, page 32.) When there is more than 1 resisting element, the relative stiffnesses are known as rigidities. See equations 65 and 66, page 71.

QUESTION 2

(a) See page 50. Definition 3 is the most common.
(b) See page 50. The method of construction, materials used, and amount of stiffness determine ductility.

(c) Ductile structures will yield prior to collapsing. This yielding removes some of the seismic energy.
(d) Local yielding will reduce the energy of oscillation. See page 44.
(e) See page 56.
(f) See page 55. (SEAOC Commentary, page 19-C)
(g) In proportion to element rigidities. See p. 70.

QUESTION 3

(a) See pages 53 and 54. Moment resisting space frames ("braced frames") do not make use of extensive joint stiffening.
(b) See page 53.
(c) See pages 50-51.
(d) See pages 63-64
(e) See pages 60-61.
(f) See pages 53-54.
(g) See page 93. This high value is justified since miscellaneous structures are not designed with the same regard for earthquake resistance as are buildings (i.e., multiple structural elements, large damping ratios, planned plastic yielding, etc.).
(h) See page 54.
(i) It doesn't. (See page 74.)

QUESTION 4

(a) See page 21.
(b) See page 21.
(c) See pages 21-22. The Richter magnitude is calculated from a measured seismograph excursion. An energy correlation is possible.
(d) Structural (load carrying) members that are not needed for stability are redundant members. Redundant detailing (extra welds, extra flange clips, extra confinement, etc.) appears to be on the increase. Redundancy is structural main members is expensive and depends on the designer.
(e) See pages 63-64.
(f) See page 50. (1) the type of construction; (2) the materials used; (3) the amount of stiffening; (4) the amount of concrete confinement; (5) the types of beam/column joints; (6) redundancy.
(g) If a building experiences torsional shear stress, the shear stress will counteract the seismic stresses on one side of the building. This is negative torsional shear stress. See page 57.
(h) See page 47.
(i) Because damping is usually small, the actual damped period is taken as the natural period. See pages 34 and 50.

QUESTION 5

(a) This is similar to example 2, page 32.

$$K = \frac{3EI}{L^3} = \frac{(3)(30EE6)(725)}{[(20)(12)]^3} = 472 \text{ LB/IN}$$

$$(472)(12) = 5664 \text{ LB/FT}$$

The mass is $\dfrac{1140}{322} = 35.4$ slugs

From eqn's 14 and 15,

$$T_{HORIZ} = \frac{2\pi}{\omega} = 2\pi\sqrt{m/K} = 2\pi\sqrt{35.4/5664} = .5 \text{ sec}$$

{more}

SEISMIC PROBLEM #5 CONTINUED

$$f_{HORIZ} = 1/T = 2.0 \text{ Hz}$$

(b) THE DEFLECTION DUE TO A COMPRESSIVE LOAD IS

$$\delta = \frac{FL}{AE}$$

IF $F=1$, THEN $K = 1/\delta$

$$K = \frac{AE}{L} = \frac{(8.4)(30 \text{ EE6})}{(20)(12)} = 1.05 \text{ EE6 LB/IN}$$

$$(1.05 \text{ EE6})(12) = 1.26 \text{ EE7 LB/FT}$$

$$T_{VERT} = 2\pi\sqrt{35.4/1.26 \text{EE7}} = 1.05 \text{ EE-2 SEC}$$

$$f_{VERT} = 1/T = 95 \text{ Hz}$$

(c) THE FLEXIBILITY OF THE OTHER ELEMENTS INCREASE THE YIELDING WITH AN APPLIED LOAD. THIS DECREASES K AND INCREASES T. THE MASS OF THE TUBE, IF INCLUDED, WOULD INCREASE THE VIBRATING MASS, ALSO INCREASING T.

(d) FROM EQN 25,

$$\ell n\left(\frac{1}{.882}\right) = .126$$

(e) FROM EQN 25

$$.126 = \frac{2\pi \xi}{\sqrt{1 - (\xi)^2}}$$

SOLVING DIRECTLY, $\xi = 2\%$

(f) FIGURE 20 CANNOT BE USED BECAUSE THIS IS FOR 5% DAMPING. FROM FIGURE 24 FOR 2% DAMPING AND 2 Hz,

$$S_a = .9g = 347.8 \text{ IN/sec}^2$$
$$S_v = 28 \text{ IN/sec}$$
$$S_d = 2.2 \text{ IN}$$

(g) SINCE $f_{VERT} > 30$ Hz,

$$S_a = .5g = 193.0 \text{ IN/sec}^2$$

$$\omega = \sqrt{1.26 \text{EE7}/35.4} = 596.6 \text{ RAD/SEC}$$

FROM EQN 46

$$S_v = \frac{193.0}{596.6} = .323 \text{ IN/sec}$$

$$S_d = \frac{193.0}{(596.6)^2} = 5.42 \text{ EE-4 IN}$$

(h) $$V_{HORIZ} = m S_a = (35.4) \text{ SLUG} \left(\frac{347.8}{12}\right) \text{ FT/sec}^2$$
$$= 1026.0 \text{ LBF}$$

$$V_{VERT} = (35.4)\left(\frac{193.0}{12}\right) = 569.4 \text{ LBF}$$

(i) $T_{IDEAL} < T_{ACTUAL}$

SMALLER VALUES OF T GIVE HIGHER ACCELERATIONS (SEE FIGURE 23). YES - CONSERVATIVE.

(j) YES BECAUSE THE FOUNDATION IS VERY STIFF ALLOWING FULL YIELDING OF THE STEEL TUBE.

QUESTION 6

(a) USE PAGE 95. ASSUME THE ROOF/WALL CONNECTIONS ARE RIGID SO THE WALLS CAN BE TREATED AS FIXED PIERS

WALL	$(h/d)_{E-W}$	$(h/d)_{N-S}$	R_{E-W}	R_{N-S}
A	14.46	.6	0	4.96
B	14.46	.2	0	16.447
C	.34	14.46	9.4	0
D	.40	14.46	7.911	0

(b) PROCEED AS IN EXAMPLE 9 (PAGE 54, STEP 2)

$$\bar{X}_R = \frac{(0)(4.96) + (80)(16.447)}{4.96 + 16.447} = 61.5$$

$$\bar{Y}_R = \frac{(0)(7.911) + (60)(9.4)}{7.911 + 9.4} = 32.6$$

(c) The wall shears are distributed in proportion to the rigidities.

(d) See pages 56-57.

(e) The 2nd story shear from wall A will have to be supported by the 2nd story floor (1st story roof) slab between the two wall A's. Failure will probably occur in this slab.

QUESTION 7

(a) Refer to equations 14 and 15. The larger m is, the smaller f will be. Concrete buildings will have the smaller frequency of vibration.

(b) See page 90, equation 1-2.

(c) See page 50.

(d) See page 38.

(e) See figure 24.

(f) See pages 63-64.

(g) See page 67, section C.

(h) See figure 26, page 52.

(i) See figure 26, page 52.

(j) See pages 62-64

QUESTION 8

Properties of A-36, TS 5x5x(5/16") tubing are:
  area: 5.52 sq. in.
  $I = 19.5$ in^4 (both I_x and I_y)
  $E = 30$ EE6 psi

Although the tubing is "rotated" 45°, the moment of inertia per tube is still 19.5 in^4 in the direction of bending. (The product of inertia is zero due to the symmetry of the tube.)

SEISMIC question 8 continued

The tops of the columns are said to be pinned - so simple cantilever curvature is assumed to prevail. The stiffness is

$$k = 3EI/L^3$$

There are 3 different column lengths:

$$K_{8'} = \frac{(3)(30\,EE6)(14.5)}{[(8)(12)]^3} = \frac{1.755\,EE9}{884736} = 1984\ \text{LB/IN}$$

$$K_{9''} = \frac{2(1.755\,EE9)}{[(9)(12)]^3} = 2786\ \text{LB/IN}\ (\text{FOR 2-9' columns})$$

$$K_{10'} = \frac{1.755\,EE9}{[(10)(12)]^3} = 1016\ \text{LB/IN}$$

THE TOTAL STIFFNESS IS

$$1984 + 2786 + 1016 = 5786\ \text{LB/IN}$$

THE SLAB MASS IS

$$\frac{(16)(16)\left(\frac{6}{12}\right)(150)}{32.2} = 596.3\ \text{SLUGS}$$

THE SEISMIC FORCE IS

$$F = ma = (596.3)(.3)(32.2) = 5760\ \text{LB}$$

THE RESISTING FORCE IN EACH TUBE IS PROPORTIONAL TO ITS RIGIDITY. THE SHORTEST TUBE WILL EXPERIENCE THE HIGHEST STRESS.

$$F_{8'} = \frac{1984}{5786}(5760) = 1975\ \text{LB}$$

THE MOMENT IS
$$M = (1975)(8)(12) = 189,600\ \text{IN-LB}$$

THE STRESS IS

$$\sigma = \frac{Mc}{I} = \frac{(189,600)\left(\frac{5}{2}\right)\sqrt{2}}{14.5} = 34,376\ \text{PSI}$$

THIS STRESS IS TOO CLOSE TO THE YIELD STRENGTH OF 36,000 PSI. USUALLY A 1.5 SAFETY FACTOR IS MAINTAINED WITH STEEL.

QUESTION 9

(a) See pages 60-61.
(b) See pages 34-35.
(c) See page 71.
(d) Plastic hinges can be designed in any member. But, since column failure is not desirable (to keep the entire structure from collapsing) , yielding of columns should not be counted on to reduce the energy of an earthquake. Only yielding of joints between columns and girders should be counted on.

(e) Special transverse reinforcement is used in regions of high compressive stress to keep the concrete from crushing. This is necessary to maintain ductility. Supplementary ties (going across the column, not around it) are required by various national and local building codes, but these supplementary ties do not confine the concrete. So, it can be argued that the confinement reinforcing is more important.
(f) After inclined cracks form at the ends of deep beams, the load is carried is a "tied arch" form which has considerable additional strength. Stirrups in the center of the beam do not carry substantial stress.
(g) Box system. See pages 3 and 53.
(h) See pages 54 and 93. If K = .67, all of the seismic load is carried by the DMRSF. If K = .80, the frame must be capable of resisting at least 25% of the seismic load.
(i) See page 90. A minimum of 25%.
(j) Rigidity is a relative stiffness. See page 71.

QUESTION 10

(a) See equation 55, page 52.
(b) See pages 53-54.
(c) See page 54.
(d) See pages 29 and 53.
(e) See page 28.
(f) See page 54.
(g) Since bleachers are constructed as braced frames, K = 1.0.
(h) Total stiffness is proportional to
$$100 + 400 + 600 = 1100$$

Load on the 1st column is
$$(100/1100) (10) = .91\ \text{kips}$$

Load on the 2nd column is
$$(400/1100) (10) = 3.64\ \text{kips}$$

Load on the 3rd column is
$$(600/1100) (10) = 5.45\ \text{kips}$$

SUBJECT INDEX